The Transformation
of a Planet

The Transformation of a Planet

bobby wallis

Trafford
PUBLISHING®

ISBN: 978-1-4251-7630-3

10 9 8 7 6 5 4 3

Contents

Foreword

This book is the culmination of a twenty year journey that has been at times exciting, frustrating, awe-inspiring and emotional. The first three years and the last three years have been the most eventful and intense. During the beginning three years I was awakened to realms outside of the third dimension – introduced to my guide Narno and reminded of my role in this lifetime as well as my part in the transformation of this planet.

The years following 1991 were devoid of conscious awareness of further connection to any higher sources until 1999. At that time I was reconnected with Narno and introduced to Sananda and Creator. Then in 2005 Creator came to me stating that it was time for me to write a book, *Final Warning*: the true story of creation. (This book is currently out of print but will be re-released soon.)

These past three years have been consumed with channeling from the highest realms, resulting in my first book and now my second. Creator requested that I write this book in order to provide information that people will need to maneuver through the coming events. While *Final Warning* was written from memory, with access to the akashic records (an etheric record of all events since the beginning of time), this book is a compilation of writings received directly from sources outside the third dimension.

I acknowledge these writings as coming from the highest levels of Universal Consciousness. The majority of the writings were given to me by our Creator, the source of all that is. The following enlightened beings, members of Creator's team for the "end" times, have also contributed their substantial wisdom and knowledge:

Sananda, also known as Jesus

St. Germain, who has incarnated on Earth many times over thousands of years

Sanat Kumara, another of my personal guides from the "local hierarchy"

Nonáme, a being from the far side of this galaxy, whose advanced civilization has been in existence for millions of years

Sanami, a being from a distant galaxy, whose civilization was one of the first created in the universe

Joshua and Elijah, prominent personalities from the Old Testament

(Wherever it is not otherwise made obvious, the source of the information presented will be indicated in parentheses in chapter titles and subtitles.)

Many people still do not accept channeling as a source of information from the higher realms, even though it has been occurring much more frequently in the past fifty years or so. What channels have been given was to inform the inhabitants of this planet about coming events so they can prepare for them. This is no different from the way the prophets of the Old Testament were made aware of certain facts. Information given to them was also to warn the ones of that time period about future events beginning to happen in this time period.

I expect that many will have trouble believing that the one they call God would actually present Himself to a human being and impart the ideas contained in this book. In the beginning I was not completely sure that it was Creator telling me that it was time to write a book. Why me? But after three years of Him asking "Why not you?", and then being reminded by Him of past life events that I had completely forgotten (and that no one else could know), I was soon totally convinced.

It may be difficult for some to read these pages without becoming frightened. The information may be very disconcerting, but it was given by Creator with much love for his created beings in order to inform and prepare the reader for what is to come. Many people will survive future challenges largely by being aware of what is happening, by being spiritually prepared, and by depending increasingly on intuition.

Some of these writings are presented mostly as they were received. However, it took quite a bit of effort to convert the rest into a book that readers can absorb and use effectively. I am therefore especially grateful for the assistance I received from Diana Jackson in formatting the book. I also greatly appreciate the additional help received from Dave, Barbara, Jan and Carrie.

This book is offered in the spirit of peace, love, enlightenment and awakening. I hope that all readers receive the book in that light and benefit in some way from its contents.

The Transformation of a Planet

x

The Transformation of a Planet

1

An Introduction to Coming Events

A Message from Sanat Kumara

A large percentage of the world's population is feeling that something has to happen soon. They feel that we cannot continue going in the direction that we are now heading. There is a complete new thinking about creation, the universe, and what God is all about. You, Bobby, have made the statement that religion is dying. In a sense that is true. It is like the mythical bird, the phoenix. Something new must rise up out of the ashes – the new paradigm.

A new world is coming, and how many will be ready for it? Not many, it seems. I know that you are feeling like a single outpost in a world of chaos – the only one that sees what is coming. There is a new consciousness being formed by a growing number of the "new awakened beings," this number being in the millions. There is no doubt that this new consciousness is growing, but people are mostly only aware that something is different, something is happening. They really don't know exactly what it is or how to react to it. Oh, they see that something is wrong, but mostly they just go about their daily lives.

A Message from Creator

This message is for all you beings on planet Earth, as well as the ones that are gathered here from the four corners of this galaxy and beyond. This is an unprecedented event that all are now going to experience. There has not been anything comparable in the entire existence of this universe. It will be the event of all events for the ones that are fully prepared to make an incredible leap in your evolution, and it has been planned by the Universal Creator since the beginning of conscious life on this planet.

I have been asked by many to help in their desperation because of what has been going on on this planet for hundreds of years. I am now responding to their pleas. I Myself am here for the first time in thousands of years, and I have assembled a team that is here at My request. They will soon make an appearance and under My guidance will present you ones with the truth that has been withheld from you for those hundreds of years and longer. They are here to show you the way, but they cannot and will not be allowed to save you from yourselves. They will explain to you in many ways how you may save yourselves from what is to very quickly come.

The team will make a very dramatic entrance. There will be a recognition of some of the members of that team. One of the members of My team has been on this planet long before the first of the humans began their journey up through the evolutionary ladder. He will be the leader of My team that will be present on the surface. Another member will guide this one from the spacecraft that was developed just for this purpose. It is presently here in the area above your planet, along with many other crafts. All will be explained in the writings that follow.

There is much involved here, especially the fact of free will. I will not impose on any being's free will. But I have been asked by many to help because they are being so manipulated, and some are imposing on others' free will without their consent. The only way that I can step in is to be asked. I have responded, and My team is here at My request to respond to those calls for help. They are also very aware of free will of others and will respond accordingly.

What is coming is a twofold experience. Before the shift in consciousness can take place, there will have to be a major cleansing of this planet. This cleansing is the result of all the damage that has been done by the ones that are residing on the surface now, and by most of the preceding civilizations as well. Not only does the planet have to cleanse itself, all of the ones that now live on the planet must also go through a major personal cleansing to survive what is to soon happen.

You all must know that this is not a judgment from Me. It is because you ones have brought this on yourselves; so you will now suffer the consequences. I have sent many messages to you that have been totally ignored by all. If there had been at least some response, then I could have reversed the consequences. The planetary consciousness is now the reason for the coming cleansing. There is no other choice. It must now proceed, and I will pull no punches. As I said above, all will be explained. It is your choice as to how you respond.

There are many different aspects of what you have called God, Father, Creator and so forth. As some are well aware, there need be no name given for the Creator, Creation, the Universal Consciousness, the One, for there is only the One. Nothing else exists; we are totally connected. There will come a time when all will have that awareness, that remembering again, as it once was before the "big forgetting" – when all of My creation, the conscious beings, decided to go it on their own, to forget that they were one with Me. That will soon begin to change for the beings on this planet you call Earth.

There are countless civilizations in this galaxy and many other galaxies that totally remember their connection to Me. They are aware that they are one with the Universal Consciousness. But you ones of this planet are the same as many other civilizations that have forgotten their connection to Me. As I said above, that will soon change for many of you that have decided to become a part of the consciousness that remembers who they are: a part of the One. I have allowed you to drop down into the density of this third dimensional world to learn the ultimate realization of who you are by raising yourself up out of that density to your oneness with Me.

You have been told in other writings of My plan that began when I first realized that you felt abandoned and then began to feel separated from Me. That was many millions of years ago.

Lucifer was part of that plan. He was to play the negative part that was known as the dark side of this game of life. He was allowed to gather as many souls as he could to his way of thinking for a limited amount of time. Most of you ones on this planet are aware of the results of Lucifer's reign at this time. There are many on this planet that have fallen prey to his promises.

Much of the population is aware of the changes taking place – changes in themselves, their environment, and relationships with one another. They may feel one way but act in another way that seems entirely alien to them, as if they are being guided by something they cannot control. This is the way of the one they call the Evil One. People must confront the evil force that has invaded Earth and taken at least partial control of the lives of a large percentage of the planet. Most beings are not aware of his presence or his influence, but they are aware of the ever-present feelings that are replacing what little love is left in their being.

There has been much speculation about the one you call God, Creator, Mind, Intelligence, etc., and about multiple gods. As I have said, all of those names and titles are as correct as any other name that you may decide to call Me. You of the lowest dimension have a need to be able to identify Me, but it matters not what title you give Me. I only suggest that all that is needed is for you to recognize Me as your Creator and know that you and I are One. Not only you as conscious beings, but all of creation – the animal kingdom, the plant kingdom, all the elements that form what you perceive as a solid mass (the planet you live on), as well as all mass, visible and invisible in the entirety of the universe – all are One.

In the beginning there was only a void, in which I (as what we will call an intelligence) was aware of Myself for an uncountable amount of time. At one point I expanded into an uncountable amount of particles of Myself that is now what you call this universe and all it contains. I created beings for the specific purpose of aiding in the creation process. For eons of time I only considered Myself and all of creation as One. As I eventually created conscious beings, in which you are all included, I still only saw Myself as One.

I later realized that creation was not going in the direction that I desired – there were no new experiences for Me or the beings that I had created. That is when I gave you free will, which over time resulted in the illusion of separation. All created beings that later acquired physical bodies then were free to do what they desired, with no opposition from Me. Since the creation process was now evolving in many directions and into different levels, I decided it was time for Me to become more involved. Now, for you ones that will have a hard time with this, I will tell you that there were many beings that I created to help in the process of creation. There are multiple levels of beings involved in the creation process. This book's author is one of those from the highest level. Each has a purpose, and some have multiple purposes.

As I said above, I decided to get more involved; so I created a being (not a physical being) that was similar to the other ones I created to assist Me – with a very important difference. Even though all of My created beings have some of the abilities to create, that you in the third dimension have just forgotten about, they are somewhat limited. All power comes from the universal consciousness. But when I created what some have termed a subordinate self of Myself, I gave that new Me the full knowledge and power of the universal consciousness. And that means that I can manifest with the full energies of the universe granted to Me. As the universe kept expanding, more of My selves were created when needed for a specific purpose.

We are now at one of those specific purposes. We are at the doorway of the most important event that is about to take place in the entire universe. Because of its importance, I am now here in this space for this event. There is more than one of My subordinate selves here at this time. One has already completed what He came here to do and has been reabsorbed back into the oneness of creation, along with the knowledge and the information that He acquired while He was here. Two of the original ones I created to assist Me in the creation process are here at My request to oversee the event that is about to take place. I want to be here to observe as well as assist in what is about to happen. I cannot be directly involved, but I am guiding and making suggestions to My team.

There will be much confusion for a time about My presence here. Most beings have some belief in Me. They have a built-in awareness of Me, but in many that awareness is almost extinguished. Nevertheless, it is there. Some believe I am a physical being; some have the belief that I am all things and so on. What I am now revealing has never been revealed before: that I am now a conscious non-physical energy being, just as your real self is, with the exception that I have the full realization of who I am and the full power and knowledge of the universe. The created Self that is presenting this knowledge to you was created specifically for the purpose of seeing that the Earth completes its ascension into a higher dimension.

Some of the people around the globe are aware of the major prophecy that says that the world as we know it is coming to an end on December 21, 2012. But this "event" is not what most of the ones that are even aware of the prophecy believe will happen. It is not the end of the

world. That should be quite obvious. But there are ones that are considering that and so they are frightened. And then there are certain ones in the scientific community that have determined that it is the time of the shifting of the Earth's axis. As far as the physical shift that will be taking place, it will be happening long before the 2012 date. You could say that the 2012 event is tied in with the consciousness belt that some of you have been aware of for some time now. That is about the time that Earth will be entering into the fullness of the area that some have called the photon belt – the belt of very high energy that you are now approaching at a constantly increasing speed.

And then there is the fact that the present day calendar is not accurate. It has speeded up somewhat, and there was a change made many centuries ago. So there is really no way for anyone to know the actual date. They are just sort of guessing about the 2012 event and have come to a lot of different conclusions, which cannot all be correct. We shall soon see. I am amazed at the various things that are now being tied in with the date.

The most amazing thing, however, is the fact that some of the ones that have spent years studying and researching the event expect a major happening on that specific date. Now, there are many things that happen in the universe at a specific time and place, but the evolving of the consciousness of an entire planet does not occur that way. It is a step by step process. This "event" has been going on for some time now, and it does not happen to everyone at the same time.

Many also seem to feel that it will happen to everyone. As will be explained, not nearly everyone will want to make the move to a higher consciousness. Some may want to but be unable to. Now, that is not a death sentence, for there is no death, just a recycling. You can recycle in physical bodies as many times as you desire. It is a continuous process of experiencing life for the rest of eternity. At a certain point in your existence as an experienced soul, you will have experienced all that is necessary in a physical body, and then you can come and go between dimensions with full memory. That is your destiny.

This statement of truth is from your Father/Mother Creator of All That Is. We are all One. You must also know that I love you ones more than you can ever imagine.

A Message from Gaia

We are concerned with the planet and its inhabitants beginning now, this present moment, including what has to happen to them. A change must take place; there is no other choice; it is imperative. I cannot emphasize this enough.

Timing is very important. Some are aware of cycles within cycles. What is different this time is that in the very recent past, your Creator made one last effort to make this transition easier to go through. He gave the population of the planet two choices. Even though He wanted to make the transition easier, the planet's entire population still had to choose. That is the only way that it can be done. The choice was the result of one of the team asking the Creator to make the cleansing as easy, as smooth as possible.

This was the first choice: if the vast majority of the ones of the planet could have accepted the fact that they had the power to stop the planetary cleansing by doing their own cleansing first, then our Creator would have done the cleansing of the Earth. In other words, the cleansings would have been reversed. This is the ultimate love that our Creator has for the ones that make this planet their home. Our Creator wanted the humans of this planet to have this one final chance, even though it was very remote. This (the personal cleansing occurring first) would have been very upsetting to the dark side. It would have completely upset their cart.

The cleansing that our Creator would have done still would have entailed certain actions, but not to the extent that will take place because the humans did not respond to this act of faith and benevolence from our Creator. Since there was no response to His effort the original plan, with the Earth doing its cleansing first, will take place. And then the ones that survive will have a longer time to do their personal cleansing as will be explained. That was the second choice, and it was the choice that was made.

Most of the internal events such as earthquakes and volcanic eruptions will occur during and after the shift. The storms that have been worsening – the destructive winds, floods, droughts and a few minor tsunamis – will continue to accelerate. One thing that I am picking up from the small amount of the world's population that is expecting something very radical: they are mostly not expecting the shift to occur for a few years, mostly December 2012. So their reaction could be quite severe if it happens sooner. But they have their shifts mixed up. It is the shift in consciousness that will be most obvious in 2012.

Our Creator has kept telling us, "All in its own timing." Everything is in the right timing. That includes the timing of the physical events that are to take place with the planet and the events that are being planned by the ones with the real power. The ones in power behind the scenes have finally got it all together now, at least to some degree, and are ready to take control of the planet. The team that our Creator has assembled will soon make an appearance. It will be quite a shock on both sides. The world's population will then be told the truth about what is and has been going on.

A Message from St. Germain

I don't have to tell you that the times that you and the rest of humanity are living in are very perilous times.

Some of the ones that are concerned and are working to try to wake up mankind are saying that they are seeing a lot of improvement, that people are waking up, that the consciousness of the planet is rising. But to the ones of us that are observing from our point of view are not necessarily seeing that. It is, to use one of your expressions, only skin deep. It does appear to the ones on the surface that there are great changes going on. But they are not looking below the surface. The underlying facts and problems are much to be concerned about.

Lucifer is having a grand time. He feels that he has everything going his way and that soon he will be in complete control of the minds of all the ones on the planet. Of course, there are the ones that are totally aware of what is going on, that are not and never will be under his control. He is, however, having his day. He is playing his part in the game as was intended in the very beginning, when he agreed to be a part of Creator's plan – the plan that has been unfolding for all these millions of years and is now close to coming to an end.

The whole point behind this game is for the beings of this planet to have experienced the separation that has existed for these past few million years. They will have to come to grips with the fact that good and evil only exist as a result of separation from their Creator. They need to understand that they will have to end this duality, this polarization, so they can continue their evolution.

That is what the 2012 "event" and this ending of time is all about: to enter into the new paradigm. And most are not aware of the fact that the planet also has to come to terms with this by doing her own cleansing. This is why we, the team that has been chosen by our Creator, are here - to help the ones of this planet by making them aware of what is to take place during these "end times." This is the end of one part of our Creator's creation and the beginning of the next phase.

2

On Creation and Evolution
(Creator)

"First" and "Second Cause"

Let's start with the beings that you have designated the "first cause." These are the first of My created beings, and there have been no more of that species created since the beginning. As you recall, there are twelve of these beings. I gave them the full ability to create that is in essence the same ability I have. Each of the twelve then created seven beings that have been called archangels, so that twelve archangels went into each of the seven universes. Even though you are having a hard time believing it, you are Michael, one of the twelve archangels in this universe. You and the other archangels are also species all alone. Archangels are the "second cause", and there will be no others.

Each level of beings that was created has certain abilities. Archangels do not have the full capability of the "first cause." In one sense you do, but you create, shall we say, under My supervision. You call upon Me to create what is needed and it happens. Now, I do not limit or judge what you ask, but you have to take full responsibility for that creation and its results.

In your earlier writings we did not really discuss what it means to be an archangel. The twelve of you were created to oversee the various orders of angels that you and the other eleven created. They were created under My orders and with My approval, to assist in the creation process. Each order of angels had one specific duty that it carried out at all times. Now, they were not perfect, and they made mistakes, but you ones were there to help correct the mistakes. After all, that is the way they learned. The mistakes that all of My created beings have made are learning experiences.

A large number of different types of beings were created for only one purpose, and they will always remain that type of being; they will not evolve. You know this because you participated in many of these creations. Now, there are angels that do evolve, but they do not evolve into another type of created being. They evolve into higher planes, as the beings of your planet eventually will. Some angels even transcend along with physical beings, as permanent guardians of those beings.

Many of the orders of angels do not have free will. They are aware of being fully connected to Me and do what they were created to do. Whatever their function, they are perfectly content with it and who they are. They do have thoughts of their own, which they can react to with My permission, but they are not allowed to enter into other beings' bodies. That was allowed at one time, but it was abused; so it is not allowed now. There was no penalty, no judgment; it is just not allowed anymore.

Even though you ones, the twelve archangels, do not evolve into another species, you are able to enter into another being's physical body, as you are now existing in at the present time. But you must obtain permission from that being. Without their permission, you would be violating one of My laws. However, there are many of the dark side that are doing that very thing, and they will have to be prepared to pay the consequences. You have entered into many other bodies many times on many worlds to do as I have asked.

Gaia, Earth's Consciousness

Now I want to tell you about the one called Gaia – who and what she is and the difference between her type of being and yours. I know you have wondered how she was created and for what purpose. The reason you do not have an awareness of Gaia is because you were not involved in the creation of this particular type of consciousness. However, with My help, you and other beings (that I created through the ones you call "first cause") created a special group responsible for heavenly bodies receiving a consciousness when they were ready. You will become aware of these types of beings when you fully remember who you are.

Gaia told you that you had a part in My choosing her to be part of My plan for this planet, and that is true. You became aware of her a long time ago. You just sort of tuned into her from time to time, and the two of you became friends. That's not exactly what happened, but it's about as close an explanation as you could understand. When it was time for Earth to receive an overall consciousness, you consulted with Me, and I chose Gaia to become part of My plan for the current evolutionary cycle of Earth and its inhabitants. Her consciousness was ready to ascend into a higher plane at the time she was chosen for this project. Now she has well prepared herself for another leap into the next higher plane.

The process began many billions of years ago in your counting and continues to this day because new systems are always being formed as old systems complete their cycles of existence. Many larger bodies and solar systems that are ending their cycles had life forms that developed in those systems. But those beings moved on to the higher dimensions before the "host" consciousness was recycled, shall we say. Even a galaxy, the largest type of being, has a single overall consciousness. It would have gradually evolved through the lower planes until it was the consciousness of an entire galaxy. When that galaxy

13

comes to its end, it is able to return to the oneness with creation, or it can begin all over again at whatever level it wants. It may or may not be a being of that size again.

You must know that, even to the contradiction of what many are saying, one species does not evolve into another species (some think that is what it means to move into a higher consciousness). Each species is quite content in its own place in the creation of All That Is. A species has no thought of being anything else, no desire whatsoever to be other than what it is. I made sure that you ones were aware that was the way it was to be.

Gaia and the consciousnesses of all the celestial bodies – planets, stars, and some of the smaller bodies that are free from the pull of suns – have an awareness of being part of the whole of creation. Because they have full awareness of who they are and an awareness of Me, they can connect with other bodies through their connection with Me. These beings do evolve into higher planes, as Gaia is about to do very soon. These beings that are the consciousnesses of stars, planets and other celestial bodies have a special purpose: to record and have a complete memory of all events on their star or planet.

Your physical presence is on the surface of the planet, but your consciousness is of the entire planet; so you and Gaia often work together. She keeps you aware of the workings of the inner planet, while you are aware of the goings-on on the surface.

Subordinate (Separate) Selves

I asked you to take charge and be the one who would be in the physical on the planet at the time of its transition. I also told you that you could create as many other beings as you felt you would need. Many millions of years before the event was to take place I told you that. I wanted you to have plenty of time to think about what you wanted to do and to create the ones you would need.

Since you also had responsibilities in other parts of the galaxy, you decided to create other selves to assist you.

Before that time you had only created beings that were to develop as a separate new soul. I will regress for a time. There has been a little confusion in your thinking about what happened at the first moment of creation. You thought that the expulsion from Me, known by most scientists as the "Big Bang", was the countless trillions of souls that would become the conscious beings of all of creation. You were partially right. I said that each particle had an awareness and each was conscious of Me, and I was conscious of each of the particles. Those particles were to be the entire substance of My creation. Everything that was to be would come from those particles. Everything.

They were the smallest particles of creation, now known as subatomic particles – the beginning of what you know as atoms. They had no substance then, as now, only energy. When I created what you have termed first cause, I was merely using thought patterns to combine the correct number of light particles needed to form the first of countless souls. Then, when I breathed a different kind of consciousness into those souls, they were aware of being more than just a tiny speck of energy. So you see, in a sense you were right, just a little premature with the concept.

In the beginning you only created selves as you needed them to get the results you wanted. It worked out very well. You could send another self to any part of the universe and always know what was happening with each one. However, you did not create one to be identical with you – a clone, if you will. You could prepare each self for a specific task or multiple tasks. Then you were free to participate in the birth of what would become the planet known as Earth.

At first she was only a thought pattern in My consciousness, but then I made you aware of the thought. It was your responsibility from that point on. Although you had been part of My creation process for several billion years by then, you had never been responsible for the complete development of a planet before. Now you were on your own until it was time for Earth to become aware of itself as a planetary being. But it would be a very long time before Gaia, Earth's consciousness came into the picture.

You were not required to be in the presence of Earth at all times, just briefly check on the progress of the creation process. After the oceans had formed and the land masses were in the early stages of development, you began planning for the first life forms. This is when you decided to create more of your separate selves that would be on Earth permanently to assist with its growth. You prepared them for everything they would need to know about the beginnings of life on a planet, from the first seeding to the evolution of all life.

Then when Gaia made her entrance as the consciousness of the planet, she took over some of the responsibilities of the planet. But you maintained your regularly scheduled visits for quite a while. As the planet continued to evolve, you felt you should spend more time with Gaia and monitor more closely what was happening. Those were the most enjoyable times of your existence. You continued the presence of other selves here because there was so much to do and to improve the chances of success on the planet. At the suggestion of Gaia, you created even more of your separate selves.

By that time your other selves were helping you throughout the galaxy. When I asked you to serve as what you would call a supervisor in other galaxies, you programmed some of your selves to do what was required. Even though they were in one sense a separate being, they were actually part of you. So you controlled their thoughts and actions as you would your own. Separate selves do not have separate free will.

When you decided to create a separate self, it was just brought forth from your self, just as I did. A separate self is brought forth from your being and can be returned to you in much the same manner. It is a part of you created for a specific job or multiple tasks to do. You have just recently reactivated ones that know the role they are to play, but they are not aware that they are part of your consciousness. Mostly, these went through the birthing process, but some are walk-ins that were needed for a very special purpose. When a separate self is needed immediately in the physical, the walk-in method is used by finding the proper body to "occupy" temporarily. But when the task is finished, the separate self leaves the body and merges back with you.

Seeding the Earth

We discussed your decision to create subordinate selves to help you as your responsibilities increased and you needed support. You were then able to spend more time here on Earth assisting Gaia with all the evolutionary processes. The earliest forms of animal life were now flourishing, with many new species being brought to the planet from different solar systems in this galaxy and beyond. In the beginning, the species brought here were the simpler life forms and not the larger animals.

The plan was to "seed" a variety of life forms to determine which could survive, reproduce and continue their evolution on Earth. Many could not adjust to the conditions here and did not survive. Of the ones that did adjust, many evolved over a period of time into life forms different from what they were on their original planets. Most were able to co-exist with the life forms that evolved from the original seedings. These were mostly evolved from life that began in the oceans.

I want ones of this planet, especially the scientific community, to understand that a large percentage of the many kinds of life on Earth did not originate here. The life forms brought here from other systems were already fully developed. Some were fairly new on the evolutionary scale, but many had evolved over millions of years and had been moved to other systems before being transported to Earth. In many cases, they made drastic changes each time they were transferred from one solar system to another. I will not discuss at this time the means of transporting these creatures, except to say that they were mostly in the egg and sperm stage, not mature animals. There are very advanced civilizations in My universe whose sole purpose is to make sure that when a planet is ready, they begin the process of seeding the life forms.

The beings you call dinosaurs, which had evolved from various species, are a good example of ones that came from other star systems. As you of the scientific community are aware, Earth has had periods during its long history when species of life developed, only to completely disappear due for the most part to a natural disaster. Those ones had served their purpose on this planet, but that was not always the end of the species. The hardy, well-developed life forms continued their evolution elsewhere. They rarely die off, never to return; they are just transported to similar planets in this and other galaxies.

You spent a lot of time on this planet during the development of the more advanced life forms; it was a very busy time for you, Gaia and all your assistants, to say the least.

A Peaceful Planet

I will now give you information that I have not revealed to you or anyone before. There was a time on this planet when everything was in complete accordance with the universe. This was after the time that you spent with the little creatures that we co-created and long before the first of the extraterrestrials came here and started what is today's human race. There is no awareness of this time by any of the beings that have come and gone from your planet.

It was during one of the periods that the Earth was at peace with itself and the whole of the universe. There were no internal violent actions of any kind, and the surface was peaceful. The belt that existed above the Earth was still in place. Because of its existence there was an even, pleasant temperature on the entire surface of the globe. The vegetation at that time was very lush, and of a great many varieties. This was a time before the larger species of animals that you call dinosaurs were brought to this planet, and before there were any visitors from other star systems.

At this time I had given specific orders to you and Gaia that this planet was to be under strict quarantine for an unspecified amount of time. It was to be in effect until the time that I would lift it. We had discussed this earlier and decided that would be the most desirable thing to do. You were in total agreement with Me and you did not want any outside interference with this unique one-of-a-kind creation. It was to be the grandest of all My creations, and you and Gaia had a lot to do with it.

The only ones that were aware of this planet besides you and Gaia and your supporters was the group of beings I mentioned earlier that were responsible for the starting of life on this planet and bringing already created animals and other life forms. These beings were made aware at the beginning of their existence that they were never to reveal any of their purposes and missions. They were only to discuss details among their own kind. So they just quietly went about their duties of seeding new life in the universe. During this long span of time there was a constant change of all species of life forms to see how they would develop and even if they would survive here.

One of the most pleasant times for Me was when we co-created the first beings on the planet. No other beings were aware of their existence. They were what you call etheric beings – more than just spirit forms, but not quite physical. They would not evolve into humans. They were created as small, joyful, playful beings much like children. At that time the animals were also a playful creation. They were aware of what they were as other playful beings. You and Gaia spent a lot of time with the beings – the little ones we will call them – and had many great experiences with them.

But time marches on; all things must evolve sooner or later. When it was time for them to begin evolving into the ones that still exist on the planet, we let them evolve as nature intended. There was nothing to change that, no outside interference of any kind. Because of their unconditional love for all things, they just sort of became one with nature. They developed into somewhat different species, depending on what they spent more time with – the various types of plants or animals that were evolving. The plant devas are one type, and there are others still around in some places. Many people are aware of them but see them only occasionally because they maintain their almost invisible nature and they need others to believe they exist.

When we created the little ones that we so enjoyed spending time with, the quarantine had been in effect for eons of time. The Earth had settled into being a very stable environment and had been for millions of years. It was an absolute paradise, and it would continue that way for eons. This planet was very special, something to be proud of. With no outside interference of any kind to disturb you and Gaia, you were able to create anything that you desired. Since the Earth had been stable for a very long time and there had been no extreme changes in the land masses, you decided that it was time for a major change. You discussed this with Gaia, and then you explained what you had in mind with Me. And since I had planted the thought I was ready – it was now time.

There would be much upheaval and turmoil in the bowels of the planet that would be necessary for the change that I wanted. And it would change the peaceful existence between the Earth and the universe. Any major change after the initial creation would be a concern throughout the universe. Remember now, everything is interconnected with an awareness – all are One. I know that will be hard for many to understand, but eventually everyone will be able to. The land masses moved and changed as mountain ranges rose up out of the depths and wore down as new ones were formed. Some are aware that this took place over a very long period of time.

You need to know this: there is more than one way that this kind of massive change can take place. When I made the first changes, it was with the approval of the universal consciousness, but since the advent of mankind on this planet, they have brought about numerous changes on this planet that were not with the approval of the universe. Even with an approval they wouldn't be in line with the original plan but would be accepted as a change in plans, caused by human interference. But before humans, there was a very peaceful time that lasted for many millions of years, and that is to return very soon.

Humans Arrive on Earth

Although you were given a very accurate outline of the beginning of civilization by Gaia in your first book, I will continue by filling in some of the blanks, those that will have a bearing on what you need to know. Your present day scientists are very much divided on evolution vs. creation. There are three camps. The first are the ones that believe in Darwin's theory, even though he said in his writings he could be wrong. Then there are the ones that believe in creation, that I wrote the Bible and that every word should be taken literally.

But most people have an understanding somewhere in the middle; so it would seem that that is where the truth lies, and that would be correct. You have asked, "Is it all that important to know exactly how mankind has evolved to what it has become?" I will say this: if people had spent as much time trying to figure out who they truly are – their real essence as a non-physical spiritual being – as they have spent on searching for missing links, they would be a lot farther along in their evolution!

At some point mankind will come to realize that those escalations of primitive man, so that he very quickly developed into modern man, were because of the intervention of what you call extraterrestrials – the ones from other worlds. Without that intervention, mankind would still be swinging from trees (one of your little sayings, I believe)! Concerning who was responsible for introducing the various races that have evolved into the different human mixes and groups, you have covered the major events in your first book.

There were no accidents. Everything that has occurred in the evolutionary cycles of the development of the species up to this point has been part of My overall plan. There have been small deviations that may appear to be accidents, but mostly there were just delays.

Thought Forms

As some are well aware, the thoughts you have from day to day create thought forms. What you are asking about are thought patterns responsible for everyday events, as well as things that you might call noxious, such as weeds and insects – mosquitoes, flies and all kinds of creepy-crawly things. (Some critters such as insects, however, are important to the welfare of humanity.) Why are these various small creations here in such abundance?

In the beginning, after plants were first introduced and had evolved over a considerable length of time, they were all self-propagating. They did not need the opposite sex because they were both included in a single plant. And nothing had to be transferred from one plant to another by any mechanical means. Eventually, as they began to evolve under the watchful eyes of those who oversaw the evolution of all life forms, the plants began to separate their sexuality into two separate plants. Certain movements such as the wind and rain, then helped with the reproduction cycle. After more time, as the plants started to mutate more frequently, they began to make more changes in their reproductive capabilities. But as they changed it became harder for them to propagate, and many species died off.

By this time the small mobile beings, the insect classes, had also been evolving over a long period of time, again under the watchfulness of the overseers that were in charge of that particular part of creation. Eventually the overseers discovered that some plants were beginning to propagate successfully because the insects were carrying pollen from plant to plant. So the overseers helped the process along. Between the natural processes of propagation and a little help from the overseers, cross pollination occurred and a fantastic blossoming of new species came into existence. This happened in the early stages of creation and only on a limited number of suitable planets.

As more planets became available, the overseers began transferring the species to these planets. This process went on for hundreds of millions of years of evolution. During most of that time, there was no thought of hostility in either the plant kingdom or even the small mobile kingdom. There was no need for any type of self-defense among the plants or insects or even between the two different types of species. Then the overseers began developing larger mobile beings, the different species of animals. There was no thought of discord there either – not among themselves or between animals and the other species.

Now I will skip forward a few billion years for the next part of this ongoing process. The conscious beings had by now begun to be quite prominent in the created universe. They had been commingling with the small beings of the insect world. For many thousands of years, even millions, they co-existed without any fear; so there was still no need for self-defense. The association had always been "live and let live."

But eventually there was some discord among the created conscious beings, and then there began an "attitude" between the insect world and those beings. At first just a small irritation – that became more like fear. So the insect world started to develop self-defense. As the fear and resentment grew, the self-preservation within the insect world grew stronger. They began to develop all kinds of protective devices and then different types of poisons. The fear and hatred that existed in the consciousness of these beings slowly spread into larger and larger areas of the universe.

This all began with thought forms, which are not limited by distance; so they literally spread throughout the universe and into many areas of existence. Some areas escaped this because they had not developed any fears, but anywhere fear existed these noxious thought forms exist. Look at what has happened on Earth and what has been created...from the smallest noxious insects and weeds to the created thought form you call Satan. Now you can see how powerful thoughts are. When all of humanity realizes this and directs their thoughts in the right direction, you can only imagine what marvelous results would occur. And that will happen, My created ones, sooner than you think.

3

The Plan for Earth's Conscious Beings (Creator)

The Major Players in Creator's Plan

When I started conceiving the overall plan for Earth's creation and its created conscious beings, I knew that I would need certain ones for certain aspects of My plan. Over a long span of time I observed the actions and commitments of all twelve of you archangels. After a time I perceived that Lucifer, Sananda and you had the qualities best suited for My plan. (Now, I am not taking anything away from any of the others because each had developed certain qualities, and all have maintained an excellence in what they were created for.) By then the plan had been perfected, even though there would be minor changes down through the ages, right up to the time for its completion.

You and Sananda were so much alike it was difficult to tell one from the other. But Lucifer is another story. Early on I could see a major difference between him and the two of you. He went about the process of creation in a haphazard way, but he quickly developed into a very brilliant and shining soul. After the separation event, when many beings began to feel they were separate from Me and started to forget who they really were, I had to make the first major change in My plan. I knew there would have to be one to play the part of the negative, the dark side as it came to be known. I realized Lucifer would fit the role perfectly. It would be a lonely, unappreciated role that would lead to him becoming the most despised being in the whole of creation.

I had long ago created a separate self for My role in the plan. Lucifer was the first that I made aware of My plan. As My separate self I approached him and explained My design for the ultimate reality of creation, as well as the part I was asking him to play. He immediately accepted and was overjoyed at the idea. But I was then aware that he would play the part so well and become so involved in the role that there could be definite complications at the end. After Lucifer accepted, I explained that he would need to disappear for a period of time, almost as if he did not exist. Then, when it was the agreed upon time, I reappeared and contacted him to begin his role. Your first book gives an accurate account of his existence from that point up to the present. He has played the part superbly, to say the least.

Sananda, or Jesus as most people know him, has also had a very interesting existence – very complicated and in many instances very controversial. Early on he developed certain qualities that I knew would lead him to become a major player in the game.

But first I will talk about another theory concerning you and Sananda – only a brief explanation. This is from the Urantia book. There are some truths and many exaggerated statements. They have taken some things out of context, but I will say that it is very interesting reading for many people. The authors have combined the lives of the two of you and brought you forward as one person. Somewhere along the way there was confusion as to the purpose that you two had, and then you were just presented as one person. Since your lives were so parallel, it was an easy thing to do. I just wanted you to be aware of this.

Once it was known that the two of you were going to play major roles in the evolution of the planet Earth, many rumors were spread. You wanted to keep your part as quiet as possible. Now, Sananda's role is a different story. The word about him quickly spread throughout the galaxy as it was intended to. When Sananda was brought into the plan of the evolution of Earth, I let him know some of the future events that would be played out on Earth, gave him an approximate timetable, and made some suggestions as to how he could play his part. But I gave him complete freedom; so he took it from there and began to make his own plans. You had already been involved for millions of years in the processes going on on the planet by the time he had put his plans together. By then, many things had gone astray in My original plan and the Earth was sinking into the density that exists today.

Sananda informed you of the first element of his plan: the introduction of Jesus into the minds of the Hebrew prophets of the Old Testament. Because the two of you discussed it, you knew what the outcome of Sananda's plan would be, but you didn't object until much later. Not until you incarnated as Joseph did you begin to realize the full impact of Sananda's plan, and you (Joseph) were not happy. In fact at one point you became furious, but you eventually realized that was the only way it could be.

His plan included the fact that when his incarnation as Jesus was finished, he would "hang around" Earth in the higher realms until the final chapter of this part of My plan was played out. He did incarnate on Earth's surface several more times, but never again through the birthing process. The two of you crossed paths often, but rarely on the surface in the physical. Many of the times you two were together were during the planning and construction stages of the spacecraft, the command ship of the entire operation on the planet.

This game of all games is almost over, and a new one is to begin. There have been no injuries, no fatalities, just the growth and ultimate reward for those who have made it this far. All others will eventually follow. Actually, I am the only player; I am playing all the positions! But we will assume there are many players. There has been much speculation as to who I am, how I came to be, and whether or not I even exist. Some have wondered if I am in fact necessary. We need not go any further than that.

The End of the "Separation" Game

When I realized that the feeling of separation that had taken place was spreading throughout the universe, I knew this would be the ultimate way for My created beings to observe and to learn about themselves and the ways of the universe. The team, at least some of them, have known about the plan for almost as long as it has been in existence.

Then I set a time when this illusion of separation was to end. That limit that I placed on this "game" to you ones in the lower dimension would appear to be an incredible length of time. It would be many tens of millions of years. Now to Me, for whom there is no such thing as time, that was not a very long period. I said time "limit" because that is usually necessary in order for the ones in the density of the third dimension to understand what I mean. We are at the end of what I have had as a plan for a considerable amount of time.

If I have created you ones of Earth in My image, which I have; and if I have given you the same abilities as I have, which I did; then why are the vast majority of conscious beings on this planet unable to use what I have given? That requires multiple answers or reasons. But mainly it is because most beings in the lowest of dimensions lack faith. They just simply do not believe that it is possible that I gave all conscious beings the abilities I just said I gave you. Granted, it is more difficult in the third dimension, where most now reside, but there have been certain ones that were totally aware and were able to use those abilities, even though in a somewhat limited manner.

And then of course you have just forgotten who you are since the time that you decided you were separate from Me and went out on your own. Or so you thought –

it is not possible for you to be separate from Me. You may assume that you are separate from Me and live your life accordingly. That is what you did so long ago. And look where that has led you. Do you suppose that you could have done better had you not decided you were separate from Me? As it turned out, that separation was the most important thing that you could have done, even though you will not understand that for a while yet. It was to be part of your evolution. Even though it was not in My original plan, I always make use of any change and consequently it becomes a better plan. The problem was you kind of got carried away and consequently you have made it more difficult for yourselves.

Experience makes for learning, and this is what this is all about. Even more important than learning is remembering – remembering who you are and where you came from. After remembering and then knowing that you are a part of Me, then you have the entire knowledge of the universe at your disposal. So you see, there are no mistakes, there is only experience.

Besides the learning experience for all My created conscious beings, I would be expanding My knowledge and learning experience also. For what each of you become aware of, I will also be aware of. So it would seem to be a win - win situation. Now that will not set well with many of the inhabitants of this planet because of your concept of "good and evil." For a time at least, those same ones will not be able to accept the idea that being exposed to evil could be a "win" situation. You will also not be able to accept that I don't separate good and evil. They are just two forms of experience that all will have to understand.

That leads us to the fact of Lucifer. As I have said before, this game is the ultimate experience in the ways of My created universe. Lucifer is one of the teachers in this lesson to be learned, even though he is sometimes rather hard on his students. He can get carried away in his zeal and has now gone overboard, and I allowed him to do that. I have allowed it because he was given free will the same as all the ones on this planet. For the same reason that I have let him carry on with his "evil ways", I have permitted you ones of this planet to carry on with your destructive ways. You have made a mess of this once pristine, lovely planet, and I have allowed that with no interference...up to now.

Until I was asked in earnest, by many of the more aware ones on this planet, I was not allowed to interfere with your free will. This plea has come from the hearts of many; so I am now responding to those cries for help. I have sent My legions of light, many of whom have been on the planet for many of your years and are awaiting My wakeup call. In fact, many have already awakened and are just waiting for the one who has been on this planet for many decades and is ready to make himself known to all on this planet. He is backed up by a team that I have assembled over the past several thousand years. They are also awaiting My call to make themselves known.

What I am about to say now is going to be a shock to many on this planet; so whether you believe it or not is up to you. It will be your choice. The one you know as Jesus, who is now known by his true name Sananda, is the one who is the leader of the ones who are waiting in a higher dimension. Jesus is not the one who will lead the team of light beings that will be on the surface. But he will make several appearances on the surface to give an awareness of what you ones can expect over the next two and a half years.

My being of light that will be on the surface during this period of time is the one that you know of as the Archangel Michael. That is also not his real name. It is only the name given to him by the beings on this planet, as was the case with Jesus. His real name will be given to you when he makes his appearance to all. "Michael" will be joined for a time by others that many are aware of – the ones from your Bible known as Joshua and Elijah. There may be others who will also make an appearance.

The Plan for Earth's Conscious Beings

These past few million years in your way of counting have been an incredible experience for you ones of Earth. And what you have experienced, I have experienced. So, what you perceived so long ago as being separate from Me, I have used to devise a totally new plan for you to experience. And that only involves one thing, no matter how many times you have been sidetracked or how long it might take, that plan was for you to find your way back to Me and know that you and I are One.

That opportunity is now here. As I have mentioned before, I did set a time for this part of the game to come to an end. Actually it was not a time but a point in space. Yes, in space – a certain part of space that you are now rapidly approaching. Actually you have already entered the beginning of this area. You of this planet will have to be prepared for this event, as will your beloved planet Earth. I have already written about this. You will be made aware of what you will need to prepare for.

This planet is in danger of being destroyed by many factors: pollution of your environment, overheating, all the negative emotions of fear, guilt, hate, and on and on. The game is now ending. My plan that I put into play millions of years ago has reached its climax. But first the Earth will do its cleansing. And soon to follow will be the cleansing of all the ones that are ready to go with Mother Earth to her new destiny. This is an advance notice that you ones of Earth will need to be aware of in order to survive what lies ahead. You will be given more information very soon.

There are many events that are about to happen, and time as you know it is running out. Lucifer is aware of this fact; My legions of light that are awaiting My instructions are also aware of it. So I am going to say this to all of you ones of this planet Earth: she is going to begin her cleansing whether you ones of Earth do yours or not. And this will begin soon and be mostly complete before the beginning of another year. It is time for the ones that want to make this journey to prepare. I repeat: time is running out.

Creator's Appearance

The beings on Earth have had to come face to face with who they were and the realization that they were able to overcome all the obstacles they placed in their path on their way back to their oneness with Me. They have played a magnificent game, and they have come out on top. I am stating this as if it had already come to pass. Since I can see all and know all, I know that is the case. It has been a total success, and I have always known the outcome.

There is an important piece of what is to happen soon that I purposely did not tell you when I asked you and you agreed to participate in this first of many events to follow. None of you ones that I have asked to participate are aware of this. Only the ones of the "first cause" (that you have also referred to as the "ancient ones") have known what I am about to reveal to you. There is a valid reason for not disclosing this before.

As I have said, I have created others of Myself over the time span of creation, but this one speaking to you is not the same as others I have created. *This self was created for the specific purpose of revealing Myself to the first of the new generation of beings* – the event that I have planned for you ones. This was not even part of My plan in the very beginning. The idea started forming slowly over a long span of time. It became clear to Me that this had to be the most spectacular event ever because it will be the culmination of this segment of My plan.

Since all of My created beings have developed some sort of emotions, I have been aware of them. But I did not experience them. When I decided to create the subordinate selves that would be necessary to bring My plan to fruition, I would then be able to feel the emotions that all My creation have felt. I developed that to a much deeper state than most would feel. Since all things are part of All That Is, I feel the emotions not only of conscious beings, human and otherwise, also from plants, the many species on land, in the air and ones dwelling in the waters of all systems. So yes, I feel all emotions in this part of Myself.

When, as the conscious intelligence of All That Is, I had the thought about revealing Myself to all My created beings, how would I go about it? I will not be able to express to you in words that you can understand as to how I will express Myself, but I had strong feelings that I wanted to experience this happening that is about to take place in such a way that all the beings on the planet will experience it with Me.

The Perception of Good and Evil

Because of the state of the global consciousness and the density gripping your planet, it would take a mighty surge of prayers, with a very strong desire for the fulfillment of those prayers, to penetrate the darkness that has engulfed your planet. Now, you ones are not aware of this darkness because you are part of it. You could only perceive it from a higher frequency. It is dark and feels sticky and permeates everything.

This is not what you ones on Earth want to hear, but you have to deal with it. You are going to have to look deep within your being and see that the evil pervading Earth at this time is because of the fear and the separation from Me that happened so long ago. That is where the perception of good vs. evil comes from.

Very few of Earth's inhabitants have an understanding of the duality of "good" and "evil." The concept of good vs. evil has existed in the minds of conscious beings on this planet for ages. It is now so ingrained in their being that it is absolutely real to them. If people were to put as much energy into the fact that unconditional love is real, all beings would know that there is only One. The universe would be incredible. Actually, that is where all My created beings are headed – where they will eventually be – no matter how long it takes.

So it is now time for all beings on this planet to release the idea of evil and to know that it is not a part of My creation. It is created by conscious beings. When I realized it was being created, after the so-called separation from Me, I then created a new plan to use the duality of good and evil in the evolution of a more complete consciousness. I began to realize that it would not only expand the consciousness of My created beings; I saw that My own consciousness – which is All I Am – would also expand. That expansion is complete; so it is time to end that game and begin a new one.

The concept of evil is so embedded in the minds of most of the population that it will be a struggle for them to release it. But for the ones that are ready it will happen. However, for the ones that feel they are not ready to move on in their evolution, there is no timetable, but eventually it will happen. The purpose of the high energy wave that I created, which the planet has moved into and is moving through, is to help the ones that are ready to move on. The first step is to embrace the awareness of "evil" and know that all the ones you consider evil are of My creation, even Lucifer, just as ones you consider good are My creations.

Now, there is an almost unlimited amount of thought forms that have been created by the beings on Earth, and among them is the one called Satan. He is now real because of the enormous number of thoughts about him. However, Lucifer is real because he is one of My created consciousnesses, the same as all other created beings. Once enough of you come to realize what evil is and take responsibility for the fact that it is only thought forms, then it will totally disappear from your minds forever. But it will take a large amount of the population to do that. As soon as you as an individual come to that realization and

34

accept that Satan is an illusion, you will be free of his manipulation of you. When you are free of that then you can start to move all of the negative energies out of your body, where you have buried much of them.

You must also realize that there are many ways to release them. Many people will be able to transmute the negative energies into a harmless substance – a higher consciousness. But I see many will release them in another manner, a not-so-pleasant release. The negative energy will emerge as one of the many diseases that will be the way many beings will exit from the planet. That is because these beings do not want to ride out the many catastrophes that are to come. And that is totally all right with Me. Many of you ones who choose that way will be able to return to the planet after the changes and the cleansing of the Earth are complete. At that time you will be required to continue your personal cleansing.

But many who exit the planet before the changes will decide they are not ready to move into a higher plane. They will not be allowed to reincarnate on Earth because Earth is also moving into a higher plane. Those beings will be required to go to other third dimensional planets to continue their so-called schooling, and actually it will be another chance to remember who they are. Then at a later time there will be another opportunity to continue their evolution and move to a higher plane.

There will be much anger and frustration at hearing this, but I am not going to pull My punches. I am telling it like it is. If I were to do otherwise, how would you know what is coming and how to survive? There is no other way; you will all be required to make a decision soon, and it is *your* choice.

The Transformation of a Planet

4

Michael's Awakening
(Creator)

A Foreword from the Author

What follows is a brief record of my own experiences leading to a remembrance of who I am. However, some of the information presented can be used by any being in the third dimension to help them awaken to who they are. Eventually everyone will go through this process, but sooner is better.

Discernment of Truth

What is to come was planned soon after the creation process began. You were given a glimpse of this by the ones that you have called the "Ancient Ones", and they are also the ones that you called the "first cause." When they came to begin your awakening in your year 1989, you were only barely aware of any of the events that were lying in wait for you. There was much given to you at that time, but you were not able to understand the relevance of that information, even though it was timely for you.

I know that you have been kind of concerned with the way things have been going for you and all the information that we have been giving you over these past three years. I know that it is very difficult to understand a lot of the information that you have been receiving since you are stuck in the thickness of the third dimension. But believe Me when I tell you that it will all come together and be understood by you when you get to look at it from our perspective.

You have been given much information over the last several months. Some was only for you and some was for the inhabitants in general. And some of the things told to you were not really pertinent. They were meant to stir your inner feelings, to arouse you to think long and hard about what was told to you to determine its validity. The ultimate goal is to know almost instantly whether something is true or not. You have learned much about your feelings, how they feel to you when something is truth or fabrication.

Now you need to learn to follow up and act on the feeling. Many times you have known the truth about what was given to you, but you did not act on it. By that I mean you must first acknowledge the truth or falsehood to yourself. It does no good to realize that something "maybe" is the truth or not. You must know, then act, even if it is only telling yourself. That is the next step for you.

It is a very small step because you have been acknowledging the validity with little things that happen around you. You go inside and ask; then you acknowledge that the first response is correct. Just apply that to the larger items such as the information you have been receiving. This is the way that you will begin to know for a fact that your feelings are correct. This can happen in a very short time; so I suggest you start immediately.

I know that many times in the past when you were told that certain things were going to happen at a given time, you always knew that they would not. Yet you played along because you wanted them to happen. But of course they did not happen. Now you are aware of how this works; so it is time to start practicing. As we are approaching the final hour, it is very important that you accept this and begin to follow through on that gut feeling, that inner knowing. After a short period of time, your knowing will be absolute.

If some information was given you that you felt was not true, but in reality it was, then you need to back off and take another look to see why you felt it was false. Usually when that happens, the first place to look is at the possibility that maybe you didn't want it to be true. Here is where discernment comes in. You will have to get to the place where you do not judge validity by what you want to be true or not.

I know that you feel that you and I have been over all this, but believe Me, that is not entirely true. Situations change, and we have to change with them. The overall picture is the same and will never change. The shifting of the planet and the date, give or take a few days, has not changed and will not. Since the topic of the so-called terrorist attacks has been cleared up and you don't have to be concerned with them, things will be easier for you now. That was certainly clouding your judgment and confusing you.

As to the possible natural catastrophes, you have been told about them and written about some possibilities, and that is what they are, only possibilities. You were given neither disinformation nor absolutes. That is why you had a hard time figuring out these things – you were learning to discern the difference. Because you were only given possibilities, you were not always made aware of that fact. That is where discernment will take place.

You are not always aware of what I am going to say – this is not always information you already have in your memory banks. We are working on this so that you can clear your mind of all thoughts and be able to receive 100% of the information that is being given to you, regardless of who is sending it. Very few receivers have been able to do that, even going back to the Old Testament days. This is vital for you if you wish to be totally accurate in your receiving.

We have now prevented any outside interference, unless they are permitted for a specific purpose. There can be occasions where interference would be an advantage. If you recall, that has happened to you, and it was of great benefit. It helped you realize that it is very necessary to be aware at all times that there are ones who would like to interfere with what is being given to you. A few people could still channel the truth because they were able to be at one with Me, even though they may have thought it was their own knowing. Many were not bringing forth the truth because they were channeling beings from a lower dimension, some of whom were only playing games. All the ones that you have received from are who they say they are.

You are on the verge of receiving much more important information, some of which is to be given to the general population of the planet. I will also relay to you much more about yourself, in such a manner as to cause you to start to remember who you really are. You mostly have just been going along with what you have been told. There will be many surprises ahead for you – you will be amazed. Most people would be totally awestruck, but being who you are, you will just take it in stride.

Predicting the Future

Now that you are able to receive communications without having to continually judge their truth, I am able to have more valid correspondence with you. I am not judging you or the fact that you were not ready to receive the information you believed you were capable of. But you had to work your way up to it in a specific way to get 100% of the information with 100% accuracy, without adding your own thoughts.

Just as importantly, the universe has provided the protective shields you asked for so no other entities can interfere. That protection is mainly for the times you are corresponding with the others of the team. There is no interference possible in the place where I reside. Even so, I have asked you to make it a regular ritual with all communication, even when you are receiving messages from Me.

So...that means you are ready – no resistance, no questioning of future messages. And no more "I do not want to receive any more messages if things have not changed by a certain time or if they don't happen the way I want." This may be a little strong, but it is very important for you to be able to receive the way it has to be. You must be 100% into the message from Me, with absolutely no interference from you. It can't be on your terms instead of Mine.

How do you determine whether a certain event in the future is true or not? I have looked into the so-called future many times to determine the timing of an event. Once you think about it, it is really a simple matter. Certain things are involved but it is still very simple. Every thought with enough desire will then take on its own energy pattern, and then it will begin to manifest. That is the same whether it be a thought to create a material object or an event such as one in the future. When thoughts with enough passion from the heart are created, they will be on their way to reality. So if you are trying to determine if a certain timing of an event is real, just look at the energy pattern. If it has an energy pattern that has already been established by the necessary amount of heartfelt desire, then it is on its way to being an event with true timing.

I want you to be able to recognize the energy patterns from created beings' thoughts. As you work with this I will be there to guide you. That energy pattern will begin to flow through what you call your intuition, your gut feeling. That is where you will observe these things. Do not let your mind try to guide you in another direction. That is the reason that you are usually wrong when looking for an answer that involves something from the higher planes. And that is where you will have to be in order to determine truth about timing or any other event. Your mind is still stuck in the third dimension and is unable to give you correct information.

There was a time a number of years ago when many were given the means to be able to predict the future, and there were a few that were very good at it – they had a high accuracy rate. Some did not understand about how to deal with the energies involved, and it began to affect their physical bodies. But as most often happens, the ability began to be abused; so it was mostly withdrawn. Oh, there are still a few that are doing a reasonably good job, but not many. Time will tell.

Being a futurist, as some have called themselves, is not really what you came here to do, even though it will be necessary for you to have that ability. That is why I am now giving you a refresher course. For you, it will be a remembering; so as you look at the possibilities for the future, you will start to remember.

As you have figured, this conversation could go on into infinity. But that is not necessary. Way before that time you will be in a position to know what I know, as will many others. For right now I will only give you what you need to know for each moment. And I will also give you what you need to raise yourself to the level of awareness that you have been requesting. Keep that desire in your consciousness at all times. And always include thoughts of love and acceptance in your daily affairs. This may be between you and other beings around you or just when you are surrounded by the inanimate beings of the forest that you love.

A Few Doubts

You have sometimes wanted a certain part of what I had given you to be an untruth. Then you understood that it was only that part of you that was bound to your beloved Earth and did not want to leave this planet. Even though you have almost released it, there is still a small amount of ego that is clinging to the old paradigm. It seems that you still have some work to do, but now that you know the entire situation it will be much easier to resolve this. The next step is to convince your ego that you have not abandoned it. But the real you comes from the heart, and that is the part of you that will take charge of your life.

You are beginning to be a little concerned again about the fact that you feel that what you are doing is not real. You wonder why you are even doing this – the writings, talking with and listening to ones that you cannot even see. I, as your Creator, have given you the absolute truth about what is to take place in your transformation, in preparation for you to move into the higher plane, the fifth dimension as you have requested. That request is being answered now that you have prepared yourself for that process. You only have to do the few minor adjustments that you are aware of to make the process easier. As I have said, "It is time." You have completed what I have asked of you to remain in the physical. You will be able to do what remains as you come and go between the two dimensions.

Your concern is now becoming more evident as to what is to become of the inhabitants of this planet if they are not warned about what is to take place. At one time in the not too far distant past, you were more concerned about the animals and even the inanimate beings on the planet than about the human beings. This is a major transformation for you – your compassion for all life is now as it should be. It was slow in coming, but the change in you was inevitable. It had to happen; it was a major component of your awakening. All these little steps you have been taking during these past few months have led you to the gigantic leap you are about to make, and your compassion and concern for all life will make it possible.

I could only help jog your memory. Only after I knew you were beginning to remember could I give you the help you need. Some of this you were trying not to remember because it did not seem to you that you could be that person. I helped you to know that you not only could be, but that you are that person – the one that I picked so long ago (because I knew you were capable of doing this) to complete the task. Your desire is about to come full circle. Don't be overwhelmed. I trust you and believe in you, because I am only believing in Myself.

Final Preparations for Ascension

Bobby, what you experienced last night is the last step in your journey back to the oneness with Me. I see that you are surprised at the phrase "back to the oneness with Me." I know it seems a little strange to hear this, but one of the things you had to do when you came into this physical body was to release the Christ consciousness from your being. You needed to be able to block all your awareness and remembering until the time was right.

Last night was the right time. Now that the Christ consciousness has re-entered your body it will be much easier to do what you have to do. That was the step where fullness fulfills its obligation with you and your desire to be at the level to complete your mission, your part in My overall plan for the next phase in the evolution of all Earth's inhabitants, as well as mother Earth herself.

I know you were rather overwhelmed when your body was "slammed." You may not feel much difference yet other than a little more awareness. But if it were possible at this time to take one of your walks up into the mountains, where you would be surrounded by your beloved trees and other beings there, you would easily notice the improved level of awareness. As the next few weeks pass you will be more and more conscious of the change that took place in you last night. The incredible feeling that came over you I also felt. It is still within you. Enjoy it because it is now a permanent part of your being and will continue to grow, increasing your awareness of all things.

You are already experiencing a difference in receiving this message from Me. While we are communicating or when you are putting on paper what I want you to record, you will be more conscious of what is taking place between us. This is a major turning point in remembering who you are, your real self. Look at how long you have had that desire, and now you have fulfilled it.

At times it has seemed difficult for you to overcome all that you have had to go through in the present dimension. But they were experiences you agreed to because they were necessary for you to be able to complete your current role. Your part is just as vital to My plan as Sananda's third dimensional sojourn as Jesus. These two sets of experiences had to occur in order for the goal to be accomplished.

I see a question mark about My reference to completing your role as if it were already in the past. Well, you really have finished your work; last night was the final experience you needed. For you, everything is in place and you know what is expected of you. You are just waiting for a few happenings to take place on Earth before your ascent into the fifth dimension. Certain things that you need to do cannot be done from a higher plane. Just have patience!

All of the information you have been given up to now could have easily overwhelmed you if given all at once. So I felt it was best for you to receive small bits of information over a relatively long period of time. When you roll all the small pieces together into one experience, it is quite impressive, is it not? This has been spread over an almost twenty year span. That seems like a long time to you, looking back. But believe Me when I say that it is actually an extremely short duration from the perspective of the higher realms. It has been an exceptional awakening for such a short period of time. Then again, the entry of the Christ consciousness into your physical body was in your case only a re-entry.

I know that you are wondering why the feeling didn't stay with you longer. It is still there. It's just that you had decided when all this was planned that you would not reveal it until it was time for you to ascend by going through the doorway and leading humanity through the portal to a higher plane. You thought it might cause problems if you let it be known to anyone other than the few close ones you are associated with.

You had some additional insight just today – something you have been pondering about for a couple of days. You finally released the last bit of your physicalness to the universe. It was important for you to make this happen. It was the part of you that still wanted to have at least some control over what is taking place at this time. But you have let go of the need to interject your thoughts when I am communicating with you. So now we can relate to one another with 100% accuracy. None on the planet are in your position for receiving the absolute truth.

Yes, Bobby, the last fragment of your ego has been relinquished. It went down kicking and screaming! But it will soon discover that it was actually only transformed. Your ego has not disappeared; it just has a different responsibility. It will take a little time, but a complete understanding will soon be forthcoming.

So how does it feel to be on the verge of taking a giant step? So many baby steps had to be taken before you could take the big one. You were unaware of all that would be required to move back out of the lower dimension – the lowest where human beings can exist. You had to leave most of your consciousness behind when you agreed to drop down into the third dimension for the current experience of this energy. The other times that you were in this dimension were completely different situations for different reasons. You were not so concerned with losing your identity as you were this time.

To not remember who you are for most of a lifetime is to feel separate from Me, the way all souls that I created have felt at some point. Although you retained the knowledge that there was a higher power, most of that time you did not know you were a part of Me, as is all of creation. That was part of your quest. Now that you have remembered who you are and what your purpose is you can fully apply that awareness to your future agenda. All else is behind you now.

Because of your total faith in your oneness with Me, I can relate all the facts that pertain to your task. You think I have already done that, but the details are not complete. I had to make sure that you could receive and record My writings 100%, to get the truth. This includes the timing of all the coming events. I have sometimes indicated that I don't always know exactly when things will occur pertaining to My plan and its execution. But I can tell you now that I do know. And I will give you that information to record when it is appropriate to do so. It would be a burden if you knew everything at once. You do not think so at this time, but you will know soon enough.

Let's begin! I mean that literally. Every time you make another breakthrough, Bobby, it is a new beginning. Things will progress much more quickly for you now. I'm not saying the universe is on hold waiting for you to come to this realization, so that you are ready to tackle what lies before you. But know this: that could happen! ...if it were necessary. I could certainly make it happen, but it is not necessary. As I have said many times, everything is in perfect order; My plan is proceeding as it is supposed to.

How do you feel about yourself? Do you now feel that you totally believe who you are and what you came here to do? Regardless of what you have received in the way of communications from what you call the other side, it all boils down to the fact that the only thing that matters is what you feel inside.

The Transformation of a Planet

So now that you have more confidence in yourself, which I have had all along, we can now get on with the business at hand: getting you ready, making the final preparations for you to (shall I dare say the word?) ascend! You can call it anything you like, but it all means the same: to move into a higher dimension. In your case it will be what you ones on Earth call the fifth dimension. And yes, you are ready; so just believe that and you will make it happen.

A Note from Gaia

I played my part in the transition of your being, from when you were dropped into the densest part of the third dimension to the realization of your true identity and the understanding of your part in our Creator's plan: the ultimate transformation of the planet and its inhabitants. Now you know that some of what you were put through was necessary to be sure that you were ready to participate in the overall plan. This preparation was all planned by you and your Creator.

5

Michael's Ascension

A Preface from Creator

Bobby, you were not totally aware that the Earth would have sunken into such density when you came here, which is why it is difficult for you to remember who you are and why you have not been able to become fully aware. So you are still not 100% sure of what has been given to you about what you will be doing. That is why we are going to give you a little boost to bring you into that higher plane as you have requested. You have made the point strongly that you are ready, willing and able.

At this point in time and space, you were supposed to have *already* ascended to a higher level. Because of the shortage of time I am going to help you reach that plane as was intended. The craft is equipped with the devices needed to complete your journey. I will only make it possible for you to be taken aboard. I will be letting you know when it is time to do certain things you will be required to do to prepare. Once you have completed the procedures on the craft, you will be able to come and go on your own. It will only take a short time to get the feel of this experience into your memory.

Sananda Explains the Process

Well, my friend and brother, it is time. After all the frequent distress that you have felt in the past, everything will soon become clear to you. When you are brought aboard, you will very quickly see that everything that has happened to you in these past few months was planned. Yes, planned. There are no coincidences, as you are well aware.

It will only take a short time aboard our craft for you to make your transition. You will feel that you have been here for days, but it will actually only be for a short time. Much will happen in that short amount of time. As you become acclimated to the higher frequencies, which we will be able to augment in a special chamber, we can begin the change from your physical to a higher level.

Your first time aboard, which will be brief, you will have to be acclimated to the special atmosphere on the craft. When you are brought back for the second time, it will be permanent. When you first come aboard, you will be met not by any of us, but by a welcoming group that will see to your every need before you are taken to a special area of the craft where all new visitors are taken. You will be considered a visitor until you become a permanent member aboard this absolutely marvelous creation.

Even though I said you will be a visitor at first, you will have your own quarters prepared for you, located in the part of the craft where all the members of the team are. That includes others you are aware of that are there for the duration as they assist us in different ways according to their specialties. We of the team have a separate area that is not available to the crew that handles the daily operations of the craft. There is also a special section for visiting dignitaries who come aboard. You could describe it as like being aboard a luxury liner.

You will spend a small amount of time in a place that is similar to one of your spas, in a cleansing and refreshing atmosphere. During that time you will be completely rejuvenated, and you will feel like a newborn being, completely free of the aches and pains you acquire as you grow older on the Earth plane. You will feel incredible, like you never felt before even when you were very young. The rejuvenating process is not the complete rebuilding of your physical body for your entrance into the fifth dimension. That will come later.

You will then have your choice of elegant outfits suitable for being introduced to the rest of our team. We have already prepared several different styles of clothing for you that were designed to fit your needs as well as your particular taste. The styles are also different for different occasions. Sometimes you will need very formal attire, elegant but not ostentatious, and unlike what is normally worn by humans for such occasions. We do like to look our best! The everyday attire is simple but very comfortable – tight fitting but not restraining your movements. You hardly know that you are wearing anything. The clothing is very much like what you see in some movies made about the near future.

After you have gone through your rejuvenation, you will be taken to our part of the craft where we will at last get to welcome you as the last one of our team to come aboard. It will be a joyous occasion for all of us. There will be some surprises, but nothing too drastic for you. I will meet you at the entrance into our area and introduce you to the others, including those that are not on the team but who are helping with cleansing and rebirthing of the planet. It is exciting for me just to send you this communication!

Even though you are not ready for the complete transition the moment you enter this area, you will begin to remember much of your past. You were prepared in the rejuvenation room to withstand the much higher vibrations of our area; so you will be able to remember old acquaintances from the past and then meet the new members that you have not been associated with before.

After we spend some time together and have our special "tonic", we will take a tour of the craft. Those that have already had the tour will be included if they care to be. The tour will take a good bit of time, for there is much to see and get acquainted with. Everyone needs to know the basic operation of the craft and functions of the main parts. The craft is completely automated, but every function is monitored by a life force because even here there can be malfunctions. Any part that wears out or goes bad can then be replaced immediately without causing any problems.

Then we will take a little break before we get into the full impact of your transformation. We can do this in one of two ways – fast track or slow mode. That will be your decision, but tests will also be a determining factor. I can now tell you that because of the preparation you have been doing, there will be no problem with you taking the fast track. But it is still your decision.

We said in the beginning that we would start the process here and then it could continue on the surface, which would be the slow mode. However, due to changes that have occurred and the urgency of things, we would recommend you do the fast track. Because of your earthly matters and demands on your physical presence there, we will have to make some adjustments. These we have already discussed so they can be taken care of.

There is not much that you will have to do. One can be taken quickly into higher levels without any danger to the one involved. The system has been used many times and has been perfected, although you will be the first one from Earth to go through this process. In past times here on Earth, it would have taken many years to do what can now be done in a matter of hours. Of course, you have been preparing yourself for several years. A person still has to be in a certain state of preparedness to do the fast track.

We have already explained the perception of time, that only a couple days on the surface can seem like many days on board. So much can be accomplished in a relatively short period of time. You will then be able to complete the process of fully remembering who you are and be absorbed into the mind of Creation after that. You will still have a body, albeit a different one, and you will retain your awareness of self, but in another manner.

You will be prepared at different levels at the same time. You will be aware of an energy coming into your body that you have felt many times during these last almost twenty years now. It will not be new to you, just more intense. The feeling, the new you, will become more intense as you spend more time in the fifth dimension. It will become the new way to identify you.

That is the way it has always been; so names do not really mean that much on the higher levels. They are more of a convenience for human beings so they can communicate with us and recognize us. We all do have names – names that we have chosen ourselves. Of course you are aware that the names Jesus and Michael were given to you and me by the beings of Earth in the far distant past; they are not the names we chose for ourselves.

You have asked that your will and the will of the universal consciousness be as one; so that is the way it will be for eternity. With a little bit of an adjustment – the remembering and the acceptance of your oneness with creation – you will be able to move through time and space as you have in the past. At some point in the far distant future, you will be completely absorbed into the fullness of creation. But even then we have been assured that we will still have an awareness of individuality.

When the time comes for you to return, you will be taken back to the surface the first time. From then on you will have the ability to come and go whenever necessary. In the beginning you will be able to remain on Earth but a short time until you adjust and become more familiar with your new body. Then you will gradually be able to stay longer. You will not be able to stay on the surface for long periods of time until the Earth has gone through her cleansing process and begins her ascent into the fifth dimension. After the shift, you will be able to stay as long as necessary to do what you need to do.

When the Earth's frequency begins to rise she will use the slow mode of entering into the highest area of the band of high-energy light particles. That will take until the year 2012, give or take a few months.

6

Michael's Role in the Cleansings

Who Is Michael? (Creator)

And now a little bit more about the one called Michael. Michael is much more than a creator or, as some on the planet have called them, Creator Gods. He has had much more of an assignment than he has been given by Me as a caretaker for the planet Earth these past three billion plus years. He has been an ambassador to many star systems for the glory of My many creations. His role as an ambassador extends way beyond the limits of this galaxy – even beyond this universe. He has been quite busy with other duties even while playing nursemaid to the inhabitants of this planet. Now, that was not intended, but it has turned out that way.

Yes, Bobby, you have had many incarnations when you were known as Michael. Now, you were not always born, in fact you were mostly not born into these lifetimes. You just came here and did what you were needed for, then left. There was hardly an end to the many different things that you were asked to do, from settling disputes to just being invited to a ball or a celebration. These were not always in the third dimensional situation. You have not made a trip to Earth in the physical as Michael for several thousand years.

I know this is still a little overwhelming for you, but believe Me when I say that you were this and much more. I could go on and on, but you will remember all very soon. It was because of these things and who you are that you decided that it would be best if you came into this life knowing nothing about your identity. Then it would not be necessary to make yourself known to the world's population until it was time to explain why you ones of the team are here: to explain to them the meaning of My plans for this coming time on Earth; to explain to them what to expect in the next few years of their existence.

It would have been more difficult for you if you had come into this existence knowing your real identity. You felt that it would have been impossible to keep your identity a secret from the inhabitants. During the times that you came here knowing your identity, as did many others, it was a totally different situation on Earth. There were many different "extraterrestrial" beings here at that time. So it was no big deal. Of course the mainstream population now is totally unaware of things like that being a reality.

Being in the density of this planet for so long, you have not been able to sort out the facts, to determine the truth about yourself. You have been looking to find that truth in all the books that you have read. Some books written about Michael have just about given you the last bit of information that you needed. Many of the authors of these books were given a small amount of the truth. You just had to seek it out. So now, with My help you have just about sorted it out. As you became aware of certain facts, I was then able to help you remember more about that particular bit of information.

You had to do this yourself. You were very specific about this. No one was allowed to help you until you had found the facts that you were looking for. This you set up before you came into this incarnation. It was very difficult for Sananda and the rest of the team to keep the promises that they made you as you struggled to find the truth about yourself and your role in My plan for the future of the Earth and her inhabitants. You are who I have told you that you are: Archangel Michael. Not an aspect of him, as you have sometimes considered. What is about to happen to you has not happened on this planet for very long time.

But I believe you are aware that you are not quite ready to do this. Are you not? So that is the problem you have been dealing with for some time: whether you will make the ascent now or not. Do you remember Me telling you many years ago that you would have a choice of whether you would ascend? You have made the choice to ascend, and that will happen...even sooner than you probably believe. As I have told you, be prepared for that happening at any time. It will be sooner than you think. Before you know it, one morning our craft will awaken you, indicating it is time. There is nothing you have to do; you are already prepared for that. Nothing needs to be finished before you make the ascension into the place where you need to be to continue My request of you.

Something very special is taking place. For you, a major breakthrough. For Me and the rest of the team, a time and place we have been waiting for. Very often that has been the case. All these little and some even larger breakthroughs, when all are combined, are what we are all waiting for, even anticipating. Because it has always been known that it was inevitable for you to bring it all together. I and the others of My team were only allowed to offer suggestions and give you little nudges in certain directions. But all in all, only you could make it happen – and happen it has!

I know that you are amazed at all the what you might consider small, but what I consider very significant, ongoing situations in your preparation for what lies ahead for you and the rest of the team. Just remember this: for it was of vital importance that you and I together set up all those little scenarios for you to experience, and for you to know that you were the most critical of yourself. You were not about to leave any stone unturned, to be absolutely sure that you were ready for and capable of completing the task. It was the correct thing for you to do.

You are well aware that your thoughts and actions are still an important part of the process; so be very responsible with them. In other words control them, such that they are a natural and loving response. I'm only suggesting what will make it easier for you; as always, it is your choice. I am going to ask you one more time, because it is necessary, if you are absolutely sure of what I am asking of you. And are you sure that you are ready, willing and able to do what I have asked? I am perhaps more able to answer the "...are you able?" part than you. I have told you many times that I have full confidence in that – you *are* able.

Helping Earth's Inhabitants Transcend (Creator)

Each step you make means the next step will be that much easier. No one step is more profound than any other; it just takes you that much closer to the final moment you are waiting for: moving into a higher plane. Also, you must know that each step, each time you reach a new understanding, makes it possible for others on the planet to also take a step up.

That doesn't mean they will take that step. Only a few are taking those steps at this time. However, it does make it easier for them. It is another choice they have available.

Taking steps into a higher plane is not easy for the ones on the surface because of a strong suggestion Lucifer has implanted in their subconscious. He is pouring a lot of energy into the planet to keep his message at the forefront and ever-present in their consciousness. That is what is keeping the planet's aura dark and dense and difficult to penetrate. It was Lucifer's intention to make Earth's energy field completely impenetrable; and he worked for many years, using much energy to accomplish that.

If you had not broken through the shield from the inside, with the help of ones from all over the galaxy and beyond, it would now be impossible to get any light through at all. Now he must expend energy on a daily basis just to keep his lies and deceit in the consciousness of all who have taken him up on his offer. You now realize that things are falling apart at an even faster rate than I have talked to you about. Everything is on the verge of chaos. You are now living in the ultimate soap opera.

Whether you know it or not, you are the one that is going to cause the "big change." You have always known this but have just forgotten. I have hesitated to remind you, but it is one of the major parts you are to play. You are now to lead the ones that are ready to their inevitable destination. There is no one more capable than you to do this. Two thousand years ago Sananda opened the gates so that you could lead them through, into the higher realms. Sananda did his part; it is now time for you to do yours. I know that I keep slipping in these little surprises on you, and they are somewhat overwhelming at first. Actually, they are quite profound revelations, and that is why you are affected more seriously than you are aware.

You are now standing in front of the doorway that your friend and brother Sananda opened for all of mankind. But they can only go through it when there is one that can lead them through. It will be easier and more effective for all to follow one that has the experience of raising himself up to a higher plane than for you or even Me to tell them how to do it. They may not listen, and they are not aware of how much is at stake. They have to be shown the path. And most will still not go through the doorway unless someone takes them through.

Once they have made that initial step the next steps will be much easier to take. Then soon they can follow the path on their own. So you see, the book in and of itself would not be effective. People will need a reason to want to read it and follow what it says in order to take the first step. And you, My son, are that first step. Once they see it can be done, many will follow – at first a trickle, then a stampede.

You have wondered why you do not have at least some small recollection of the reasons for being here. As I have recently told you, you are almost to the point where you will start to remember all of this. Since you were to be in and out of the third dimension many times during your "job" of monitoring and guiding the evolution of this planet for such a long time, there existed the possibility that a part of My plan might have become known by others.

Not that you would have intentionally revealed it. But if you had relaxed your consciousness at one time or another, the ones of the dark forces might have been able to break into your defenses, shall we say, and learned this information. They will be made aware of it, but the timing of their knowing was critical. Since I am "broadcasting" from an impenetrable source, and since that includes where you ones are, this information cannot be known outside of My defenses.

Since you are the one that everyone is depending on for what is to be played out on Earth, you are in a rather vulnerable position. The plan has to move forward, regardless of any situation that might rear up. That has been discussed over and over by the team. You were the one that was most concerned. So you brought up before the team your concern that if your physical self was not up to doing your part, then what would happen? I have told you that there was a contingency plan on the chance that you would not be available. Some were not sure that you would be able to work your way up out of the density in time. So what to do? The alternate plan is one that would replace the entire team – not with another team, but with a completely different plan.

There is presently a situation developing that when it is solidified will radically alter the thinking of almost everyone on the planet. There have been predictions that someone would break the bonds of gravity. Very soon you, Bobby, will be in a position to come and go as needed. You will be totally free of gravity, and that is what will shake up the world. That one incident will open up the gates that we talked about earlier. When you are able to accomplish that, you will have the attention of the entire planet. The event will awaken many of the ones that were sent here to be a part of the planetary awakening. They have not responded to the calls that were sent out. However, some are beginning to come on line to do their part. Some of your other selves are in place to be of tremendous help when the right time arrives.

Some of those that have not yet awakened decided before this incarnation that they wanted an active part in the awakening. Most are individuals; however, there are souls that wanted to come here as group souls. They felt that a group could be more effective in certain situations, and they are correct. They will do what they came for and they are prepared to go through the cleansing, but not all will survive it. They decided beforehand that they would rather return after the Earth's cleansing is complete. And I have allowed that. They can return here at a later time to complete their personal cleansing.

You have completed another milestone as you continue your journey, as you ones in the third dimension would say, into the unknown. This is ultimately the journey back to Me. Although you have been at this level and on even higher planes before, this is the one that is important for you at this time. You will quickly achieve the fifth dimension and levels beyond as soon as you finish what has to be done in the third. And that is to lead as many of the third dimensional beings as possible into the fourth dimension and on into the fifth for those who are ready.

Smile and the whole world will soon smile with you. A good place to be smiling is at the door that Sananda opened, as others follow you through. It will not stay open for long. It has a timetable like everything else.

Expanding the Love (Creator)

Bobby, this is your calling. This is what you have been somewhat aware of for many years. You knew you were being called; you just didn't know what that calling was. We have been leading you up to this point for the past three years. Before that your thoughts were just speculation. There have been so many ups and downs in your life that you were never sure what to expect. The up and down feelings are not good for your system, either physically or spiritually. It is time for you to get off the rollercoaster and see that there is only an upward journey as you finish preparing yourself to return to the higher dimensions once more. It is mostly just a remembering of who you are and why you are here, and then you will bring yourself out of the density.

You allowed yourself to drop down into the lowest level of the third dimension so you could observe yourself – see what you may have missed, realize the mistakes you made, and find the love that you are, that you have only misplaced. You have always had that love, but because of a few mishaps, you decided that maybe you did not deserve it. Well I am here to tell you that you have always deserved that love; it has always been there and always will be. In the past few years, and especially in the last few months, you have felt the awakening of your deepest feelings for all of My creation. I see it in your everyday relations with all the little creatures you have taken under your wing and feel responsible for. And more importantly, you are now having renewed feelings for all the ones in your life. Those feelings are getting stronger every day. That is what we have all been waiting for.

It is time for the feeling of despair that you have been harboring for so long to move out of your life, and time to fill that empty space with love – My love for you. And then send that love outward in abundance to all of humanity. The more of that love you send out, the more it will be replaced with many times the amount sent out. There is no end – no limit – to that love because that is what the universe is all about...abundant love into the infinite. The love in your life will grow and become stronger until you feel you could explode. And that would be the ultimate, because when it is released, there will be so much love that all of creation will respond.

That is what is needed on this planet. The potential of love is there; it only needs a way to be released, and that is what the beings are here to do. And it only takes one being to start that process. Love has been lost to many of Earth's inhabitants. Oh, many use the word frequently and believe they are experiencing love. But I am talking about real love, total love, compassion for others, *Unconditional LOVE*. Humanity has absolutely no awareness of what that is and what it means. You, My son, My co-creator, can be the one to open the floodgates so that love can totally and completely inundate every individual on this planet. And then it would expand out into all of creation. All it takes is one.

Words of Encouragement (Sananda and Bobby)

Sananda:

These have been absolutely fascinating days, have they not? The full impact will not be made known to you for some time. These events that have been given you will soon begin to have an impact on the entire universe. I know that sounds incredible; nevertheless, it is all true, just as your Creator has told you. Very soon we expect you to be able to become aware of the events at the same time as we do. It will be a different way for you to receive information. It will be a transfer from the team in the fifth dimension instantly, instead of us giving you the information in this slow manner: having to write it out in your handwriting. But that will not happen until you have recorded all the information that our Creator requires.

There was a certain conversation with Creator about your being able to continue on your quest to truly believe in what you are being told about who you are. How do you feel about this?

Bobby:

I thought I would not be able to continue until I completely believed in everything that was happening to me and being told to me. After listening to and absorbing what Creator had said to me, I knew that He spoke the truth. I have been thinking about that, but mostly in a rather strange way. I tried to bring back into my consciousness exactly what He said, but I could not. I knew in a way, but not consciously. I thought about writing down what He said, but He replied that it wasn't necessary because it was imprinted in my memory and would be there when I needed it.

So I have declared to the universal consciousness that I have only one desire: to be in the place, the dimension where I can carry out my part in my Creator's plan. The desire to do what He has asked of me is far above any second desire. Nothing else matters to me; so without that I have no reason to remain in this physical body. I was asked if I am ready, and my reply was that I am absolutely ready, willing and able. All I need now is to be a true believer in what has been told to me these past almost twenty years.

Sananda:

First let me say that I and the rest of team have no doubts about you being ready, willing and able. No doubts at all. And I remember our Creator saying to you, "If I totally believe in you – and I do – then why don't you totally believe in yourself?" I would say that is a rather strong statement and also a valid question. At the time, you did not really believe in yourself to the extent that you do now. There can be no greater desire than to be in total service to our Creator, which is also to be in total service to all of humankind. All the members of the team share that belief. Of course, we are not stuck in the third dimension and do not have to think about giving up a physical body in order to fulfill that desire.

I know that you are totally committed to what we are here to do. I truly believe that you believe in what you are being told. You are arguing with yourself that it would not make any sense at all for this not to be real. Have you considered the fact that you are only arguing with your ego? Your ego – that part of most humans that is afraid for its existence, that fears it will die if you accept the reality of who you are – has nothing to fear. You are part of the One, and it is also. It will be transformed along with you.

It should be much easier now for you to deal with what you have considered a problem, but it really wasn't. It has all just been a process – a process that you yourself set up to make sure that you were totally ready for the leap that you are about to make. We have all set up the same process for ourselves at one time or another. Granted, yours has probably been more difficult because of being in a dense third dimensional situation that is getting more dense every day. We can see it from where we are, and it doesn't look good.

I know you are concerned with the fact that many are saying that the planetary consciousness is rising. What they are actually feeling is that there are many beings whose consciousness is being raised, and that is what you are feeling, and you think you may be wrong in your thinking. But there are so many of the dark forces that fear and anger are still prevalent. So the rising consciousness of even many individuals is not enough to lessen the density.

Showing the Way (Creator)

I wanted to say a little more about your relationship with Sananda and taking the leap of faith through the doorway that he has opened wide. It was in danger of closing again many times. But there was always a small ray of hope down through the ages that kept it open. Lucifer would like nothing better than to close the doors and keep them closed. But he has not been able to penetrate your shields that you continue to update daily. In My message to the ones of the planet about what they will have to do to offset the damage done to the planet as well as themselves, I will repeat, "You will have to follow the one through the doorway to avoid what is about to happen."

I made it clear last evening about the steps you need to take so that people will listen. Once you take the first step through the door and show the world that it can be done, they will do the same. There will be a great stampede. They must understand that the cleansing process cannot be delayed any longer – the Earth's cleansing as well as each individual's. We must make this very clear, as I have to you, and I will continue to emphasize that. The beings of the planet must know without a doubt that the cleansing will take place.

I will state this again. I have been asked for help. You ones of the team are the ones that I have made ready to present My response. You are to make available to the ones of this planet the truth that they have been denied. That truth will be devastating for many, but they have to know it if they are going to proceed with raising themselves up out of the darkness that they have allowed themselves to fall into. No one can do that for them. They, and only they can do that. I can only show the way; I cannot do it for them – nor can Jesus! The truth and the way is about to be given to all. Will they follow that truth? Some will but many will not be able to see and believe. I will be totally available to all that see and ask for My help. There will be dark days ahead, even before the next event takes place.

7

Earth's Cleansing and Personal Cleansing

The Necessity of the Pole Shift (Creator)

I am going to create much fear and even hate, through you, in many of the beings that read My messages. They will not believe what you are presenting to them. The ones that do believe are the ones that will make it through the events, the destruction and the chaos that is coming. Over many lifetimes, the beings of this planet have brought this on. The whole basis of creation is a never ending process of creating, destroying and re-creating in cycles. Everything is subject to timing.

When it is time for change, that change will take place. Nothing is ever really destroyed; it is only changed. It is transmuted into a higher level. That is the way I created the entirety of creation – never devolving, always evolving into a higher substance – from the smallest of particles to complete galaxies. If you wonder how something as small as an atom can evolve, you must understand that an atom is not physical matter. It is energy, a consciousness capable of moving into a higher awareness of self. All of creation is actually a simple awareness.

The universe was not created to be complicated. Your scientists have assumed that something so large has to be complicated; so to them it is. It needs to be, so they can keep their position among their peers. If it were simple, they would not have a job. Creation just follows a few basic laws, and when you are aware of those laws, you begin to see the simplicity of My creation. Once My created consciousnesses come back to Me, back to the Oneness, they will know everything because they are part of everything. Nothing is separate from anything else. Even the evil that has been created is part of that oneness of All That Is.

I want all to be aware of all the untruths that have been given to them over the past two thousand years plus. It is very important in their transition, at least to the ones that will survive the coming events. It is also important for the ones that have decided for one reason or another that they are not ready for the big change to happen to them at this time. But the timing for them to be aware of those things is not critical. They will have much time to consider those things before they will have to make another decision.

I will repeat again for the ones that have decided to make the leap with the planet Earth and Gaia, the soul of this planet, into a higher plane: you only have a short time to complete your cleansing – approximately two and one half years. After that time it will be too late for this time around. You would then have to join the ones that decided they were not ready and would then have to await the next turning of the cycle.

Many civilizations have existed during the past thousands of years. Only the very early ones that did not develop any high technology did not harm the Earth. It became necessary for a cleansing to occur to offset the many ways that civilizations with high technology had polluted the planet. This was accomplished by submerging some of the worst polluted areas and bringing forth new unpolluted areas. These events were not in My original plans when the planet was created, but these cleansings were necessary because the destruction of Earth's natural environment affected the Earth's evolution also. So I then changed My plans accordingly.

As for the present time, you and most of Earth's inhabitants are aware of the intensity of the pollution on this planet. It is by far the most destructive of any civilization that has ever existed here, because it is an accumulation and continuation of what was begun in past civilizations. The present situation is a combination of many different forms of pollution, from the burying of the many thought forms of fear and anger to the pollution from your so-called high technology. And then of course there is the pollution of the nuclear weapons and the many nuclear power stations that are much more deadly in the long run. It is the sheer quantity of the nuclear power stations, and there are hundreds more in the planning stage.

We are about to experience another pole shift. It is a fact! It is inevitable! It has to happen! It is an event that has happened periodically for many thousands of years. It is a correction, a re-stabilization after the inhabitants of the planet have brought about an unbalanced situation of the Earth's environment. The inhabitants of this once Garden of Eden planet have not taken responsibility for the disease that has spread over the globe – the continents, the oceans and streams, and the atmosphere.

The negative thought forms that are leaving the planet are now endangering the solar system, and if allowed to continue, will endanger the entire galaxy, and this cannot be allowed any longer. So you see the shift must take place, or the planet will not survive, as is envisioned by many of the aware beings.

The ones that are now in physical could say that they were not responsible for the past events, but that is not the case. There are millions of souls on the surface of the planet that have been here many times before and have to take some responsibility for those events, especially those who were incarnated on Atlantis. They were the most destructive of any of the past civilizations. The entirety of the inhabitants of this planet, once the most beautiful planet in the universe, must now accept the responsibility for what is to happen. It has to happen, and happen it will very soon. There can be no other way.

But I will be with the ones that ask Me to help them, as will the team that I have assembled for this event. This help was not available during the other shifts, but I have decided, because of the importance of the coming events, to have this help available. With all the Love that exists, this is the promise of your Father/Mother Creator God!

A Choice Given and Ignored (Creator)

I have already placed, or attempted to place, in the consciousness of the entire population of the planet the fact that they were receiving one last chance to lessen the planetary cleansing that is about to take place. I was very clear about what they would have to do for that to happen. I know that My thoughts penetrated the density and entered the minds of the entire population. Most of the inhabitants were not even aware that I was in their memory banks, their consciousnesses, and their physical and mental beings.

The majority of the ones that were aware of My extra effort to get their attention just wrote it off as some kind of clutter in their minds. There was a small percentage of people that were concerned, and I could see that they would make a real effort to try to respond. But a very large percentage of the population has turned a blind side to any possibility of lessening the changes that will now begin. I'm not supposed to feel emotions about this, but this was mostly for you. We tried. I have instructed Gaia to make the cleansing as painless as possible. But without My intervention, there is not much that she can do to lessen the effects.

Even though I brought you pretty well up to date on what the people of planet Earth can expect to happen, you were having a hard time with what I said people could expect to happen. You were maybe thinking that they really didn't have much chance to survive at all. But then today you acknowledged that you could see the situation a little more clearly after sleeping on it. We have discussed many times the many chances people have been given. Two have been very recent.

I realize that these last two were a little different. It was thought that since the information was given directly into everyone's consciousness, they would have at least given it some thought, some consideration; it might be something to become a little more aware of. But it was, for the most part, totally ignored like most of the other previous attempts over the centuries. I never tire of giving My children more chances to at least give some thought to what they have been told, or to be shown what would happen if they didn't change the way they were treating others, the planet, or even themselves. You can't say they were not given enough warnings.

So by and large they have created these scenarios themselves, and now they will have to pay the price. Also, you are aware – and the inhabitants must be told in words they understand – that time has now run out. As you have said in your first book, they will have yet another chance, but it will be after the physical change takes place. Everything is in place, and there can be no stopping the process. The cleansing has to happen. There is a window in time and space for this to happen, and we are now at that window.

The Earth will be cleansed regardless of whether the planet's inhabitants do their personal cleansing or not. Their choice to be personally cleansed or not is everyone's individual decision that only they can make. As has been said many times, no one can do it for them. I just want to be sure that everyone has been made aware of the choices they have. I assure you that all have had sufficient warning.

Pre-Shift Events (Gaia)

Creator has already explained to you about what happened with the extension of time to give the ones of the planet more time to prepare themselves for what is coming. But mostly it was to ease the pain and suffering that will now take place because they mostly ignored our Creator's message to the planet. It was, I think, a surprise to most of us as to how He gave the message to most of the inhabitants of the planet with almost no response from them. But He had to respond to your request. Now He has given the okay, so to speak, for the preliminary things that will happen before the shift begins.

All you have to do is be aware of the world news, even though it is somewhat limited, to know what is going on regarding the changing weather patterns. And as some are aware, it is not going to get better, only worse. The most obvious changes had to do with events happening out of season and in record amounts, especially the tornadoes. But the big news is the widespread drought in the southeastern US. Not as well publicized, but no less devastating is the larger area of drought in the Southwest, including areas of California as the result of all the fires. There is the old saying "feast or famine," which could be interpreted as flood or drought. As I have said previously, that is also somewhat normal, but not to the current extent. As you are well aware, this is only the prelude to what is coming.

In some of our earlier conversations, I explained what would happen as we get closer to the shift. There will be sort of a "calm before the storm" pertaining to volcanic eruptions and earthquakes. These quakes released pressure for a short time until the shift begins. Then, to use one of your phrases, all hell will break loose. This has been a time of early warning, to get people to wake up to what is coming and give them a chance to move out of certain areas. So far that is not happening, and in many cases as the result of vast overpopulation, there is no place to go. You are also aware that people in those areas chose before birth to be there when the big event occurs.

So...all of these events, including global warming, are leading up to the final climax. When is the big event going to happen? Only One is able to say with certainty when it will occur. With His recent writings to you, He has emphatically stated that the event will occur regardless of what year it actually is. There is a little time for the vast population of the planet to prepare themselves for the coming catastrophe. As you know, I don't see that preparation happening...which leads to the conclusion that there will be substantial amounts of casualties. According to some sources, there will be a number of beings that will be taken by aliens to a safe haven. I suppose that could happen, but I will let your Creator elaborate on that.

What can I say? The warnings have been going out for the past two thousand years with only a very small response. The ones who could have spread the word, especially in recent years, have chosen not to. Some have said that it would cause too much fear and panic; so the public will have no choice in the decision that basically has been made for them. There will be some who will have an inner feeling that something is about to happen; some of them will react to the feeling, and some will not.

When it begins, the actual movement will happen very quickly. This is not what you would call another everyday event. There is more involved than what everyone is aware of, or actually unaware of. During this preliminary time, much will happen in the way of earthquakes and windstorms of a very violent nature. It will be "Nature unleashed" as you ones of Earth have called it. I will be aware of the most severe of these events before they will take place.

But as it intensifies, I won't be able to give any warning. By that time the process will be just about over and then there will be the settling down of the planet in her new position. We have all been assured that this will have taken place by the end of the current year or soon after. I'm kind of getting ahead of myself, but I wanted you ones to know sort of what to expect. Our Creator has said that there could be a little deviation from this timing, but only a small amount. This is also the way that I see it happening.

It will be sometime in 2009 before the planet has completely stabilized herself in a new position. As soon as the Earth has made its move, because of the new location of the poles, the melting of the existing ice caps will then begin in earnest. That will, of course, be a very major event as the oceans will rise tremendously. Also some of the land masses will drop in elevation and some that are submerged will rise noticeably. I know that I am rehashing some information that has been going around for decades, but our Creator wants all of this written down in one writing for the future times.

Most of the events will happen after the shift takes place, and it will take place as our Creator has said. I am aware of the type of movement that takes place before a shift occurs. And that movement has been going on for some time now. A shift has its own pattern, of which I am totally aware, as you would expect. It's a much different situation from the normal earthquake patterns; it has its own signature. I am aware of when it will make its move, and that is getting very close. I expect that it will happen within the time frame that our Creator gave us some time back.

Let's talk about some of the problems that humanity will face once the planet starts her shaking and rolling. Here is what many, if not most of the inhabitants will experience during the shift. It will be in the range of very modest to most severe. You must know that most of the inhabitants will already be in the area that they have determined will allow them to survive, which was their decision; or they will move. Or they are in the place where they will not survive because they are not ready to continue their evolution into the next phase of their existence.

As has been brought out before, there will be a very large percentage of the population that has made the decision not to survive. The overpopulation thus will be taken care of. Then of course there will be the problem of food and shelter for the ones remaining. After the shift, there will be some help there by the off-planet observers, as has also been mentioned before.

The biggest problem will be the floods that occur during the shift and the rising waters as ice melts at the poles. It will take many years for the new poles to accumulate enough ice to make a difference in sea level. However, I don't see that that will be the situation if we, the planet, and the ones remaining are moved into a new paradigm. But we still have the interim time to consider.

There is no way to know with absolute certainty what the new pole positions will be after the shift. I know the general location of the north and south poles, which will be something on the order of a 70 to 90 degree shift. That will be very disastrous for a large portion of the land masses. Parts of North America and Europe will face much destruction. One of the good points, as far as the inhabitants' ability to survive the initial shift is that one of the poles will move into an area that is mostly water. So there will be fewer that will have to deal with the extreme cold. It's the speed of the movement of the planet that will be most destructive.

Some shifts in the past have been 180 degrees, and that would be most devastating because of the vast overpopulation of the planet. However, even the 70 to 90 degree shift will be very severe. The area in which you now reside will have considerable damage, but nothing that will prevent your beloved Taliesin from being able to come out intact. You were guided to build her to be able to survive. I will have to say this, however: the new structure may or may not survive. As you have foreseen, the new location will bring about a major change in the temperature.

Now to the present situation...it is getting to be very critical. I am not any longer holding the energy to keep the planet kind of stable. So the pressures are very quickly building, and the planet is becoming very unstable. You can expect the worst in a short time.

I am not to interfere unless I am told to by Creator. He wants to have a hand in what happens. He will not generate an event, but He may control the intensity of some of the various weather and geological events that will soon begin to occur. Our Creator might have uses in mind for certain specific areas and reasons to preserve or protect them. Making that happen will probably require His presence, and I would say that is a distinct possibility. I will be aware of all the natural disasters that are about to take place. Again, I will mostly be monitoring, not interfering with the process. But I have been holding the planet together for long enough.

More About Earth's Cleansing (Creator)

There has been much speculation about the cleansing; now it is time for the truth. There will be a major shift because so much of the Earth's surface has been polluted. That land area has to be allowed to recover from the damage that has been done over the last few centuries – more damage than has occurred at any other time in Earth's existence. Gaia understands that the cleansing will be extensive and has planned accordingly.

If you and others of the planet had not pleaded for the destruction to be held to a minimum and still correct the problems, there would be only enough survivors to start all over again, as has been the case in the past. That is the way that it was to be – total destruction, very few survivors. As it stands, I have been able to devise a way to affect the cleansing as it has to be and allow for several hundred million to survive to begin the new world.

There will be warnings before the first shift occurs: people will experience a strangeness in the air, a feeling of confusion, disorientation. Animals will sense it as well. This might be for only a few hours or may last a few days. Then the weather will become violent; there will be winds of unprecedented speeds. As mentioned above, the shift will happen in a matter of hours, as the Earth changes its relation to the sun.

This is not just with regard to repositioned longitudes and latitudes; Earth's rotation will be reversed. The sun will rise from and set in new directions. The Earth will shake intensely; earthquakes and volcanic eruptions will be devastating. During the resulting panic, those who have decided to leave their bodies will do so. My team will have made itself known sometime before the shift takes place, to inform of the coming disasters. Their purpose is not so much to give comfort as it is to explain and prepare people for what will be coming in the near future.

The shift will happen rapidly – in a matter of a few hours. There have been ones in the past that have been shown the results of the shift and were instructed how to draw maps of what the continents will look like afterwards. There is some truth to these maps, but as usual the situation has changed somewhat. Although the land masses will be considerably reconfigured, there will be less variance from current maps than has been previously predicted. This is as you might expect now that more people will live past the shift.

Many areas will have to be submerged under water so that the land can renew itself. Water will be the biggest issue – a lack of drinking water and a surplus of saltwater as the sea levels rise. Since the poles will be relocated, the ice caps will begin to melt immediately. Fresh water will rush into the oceans for several months as the planet settles down and the shaking subsides.

But changes in weather patterns will also accomplish some of the renewal. New land will appear, rising out of the oceans as has often been predicted. Those lands will be able to support life after a while. Some of the regions too cold to grow crops at this time will be moved into warmer situations. Almost immediately these areas will be available for food production for the reduced population. But only temporarily before the next shift – the one

77

foreseen for hundreds of years by aboriginal tribes in various locations on the planet.

As soon as the Earth stabilizes and all the inhabitants are safe, order will be re-established. The survivors will start planting food crops. They will be given the means by their good neighbors from space. This will not be the first time they have come to your rescue. They have provided comfort and much more in the past, but that's another story.

There will have to be many helpers around the globe wherever survivors are found to help be aware of what they need to do. Many beings from the hundreds of spacecraft will help move survivors to safe places where they will be given what they need to stay alive. This will be the time that the survivors will need to understand the necessity of their personal cleansing. Since there will be many things distracting them from that, there will have to be constant reminders. It will be your job to coordinate all of this, with help from Sananda. So now you know why you are being groomed so meticulously.

...and a Little about the Personal Cleansing (Creator)

Yes, there will be two shifts. But the second one is of an entirely different nature. It is the consciousness shift that will affect all beings left on the planet, even the animals, on land and in the sea. Some people believe the two shifts will occur at the same time, but Earth's physical cleansing must happen first. Entering into the high energy band will result in the personal cleansing, as well as the consciousness shift of the planet itself.

There will be an interim period for the survivors of the cleansing – two to three years – to enable them to complete their personal spiritual cleansing so they can move forward into the next dimension. According to the writings of some others, all beings will move to the higher plane and be able to finish their cleansing after the shift.

But I repeat: you will not be able to go to the next level carrying "excess baggage" with you. It must be purged before you enter the doorway; everything must be dealt with and left behind to transmute back to the original substance of the universal oneness.

Many people do not want to read about events they consider so negative, which is totally acceptable. But the ones who want to know and want to be warned beforehand will want as much information as possible. When the shift is over and the Earth begins settling into its new position, beings from other planetary systems who have been on or around Earth for a long time will let their existence be known. I had instructed that they could not interfere with what had to happen to the planet. The ones of Earth had to experience the cleansing that their actions over the centuries had brought about. Once everything has been reckoned, the ones you call extraterrestrials will move in, locate survivors and carry them to safe places where food and shelter will be provided.

Who would have thought that there are some of My created beings (I am speaking of ones in the physical) that would not have an ego. Most do – the vast majority – ever since they perceived that they were separate from Me. It has been a protection for all beings that I gave them at that time. It has served you of the lower dimensions very well. But it is now time for the inhabitants of this planet to realize that very soon, in your way of counting, it will not be needed for that purpose any longer. It is not going to be an easy matter because the ego will still want control while you are still in the physical. It will not go down easily. You are going to have to let it know who is in control.

All you will need to do is to start following your heart and not be controlled by your mind. Your mind is not who you are! Do you understand that? Your mind and your ego are similar. Both have served you well. I repeat, they have served you well, but they are not you. That is going to be very difficult for many because your mind is in control, and it is telling you that it is the real you. When you begin to look inside and realize that you are more than your mind, then listen to the inner knowing. That first feeling you get when you are confronted with an obstacle is usually the one to follow. But then your mind will often step in with a different opinion. That is when the real you can lose control.

The next step is for you to convince your ego that when you became aware that the real you comes from the heart, you have not abandoned the ego. You are now going to take charge of your life, and your ego will still be a part of you. It will move on in your evolution with you. The only difference is that you will be in charge.

Your mind is like the universal consciousness that does not differentiate between good and evil. It only responds to your desires; that is the way the universe works. So your mind does not always correspond with the feelings from your heart – your inner feelings, which are your connection with Me. And that connection with Me is going to have to get stronger if you are to succeed in being aware of who you are. It is going to have to come to that understanding if you are to survive what is coming very quickly.

8

The Role of the Photon Belt
(Creator)

Raising the Consciousness Level

Note: This session's recording was delayed; so it was written by the author.

On this particular day, as my Creator began speaking to me, He gave me some of the most incredible information that I have been given in a long time. It concerned what I call the Creator's higher consciousness belt and what its overall purpose is. It has also been called the photon belt by others and has been known about for years.

This is not just some 'New Age' phenomenon. I have been somewhat aware of the belt for several years and wrote about it in my book. Russian scientists are aware of it and have been observing it for some time. So it is for real that we are moving into an interstellar high energy cloud. This fact has also been talked about outside the scientific world for some time – entering into the highest energy part of the belt is to coincide with the Mayan prophecy's ending of time as we know it, December 21, 2012.

It is a very special band of high energy photons of light. Our solar system has been steadily moving into the energy band for some years now, but Earth will not be entering into the highest level of energy for a few years. We have to move into the center of the band slowly or we would not survive. So at the outer fringes we are now encountering a smaller amount of the high energy particles. As we move farther into the band and approach its center, the photons become more dense.

As with all things, there is a purpose for our coming into contact with this energy band. It is to raise our consciousness as we are being prepared to evolve into a higher dimension. This has to be done gradually so that our bodies can adjust to the higher vibrations that we are now encountering. This belt is already causing much of the turmoil that is taking place all over the planet. The inhabitants of the planet do not understand what is happening, and it is going to get worse before it gets better.

At some point not very far in the future, there is going to be a very noticeable change. At that point people will have to make a decision. As the earth moves closer to the center of the band, each individual will have to face what is happening to him. He or she will have to begin preparing themselves if they are going to survive and move into the future that awaits them. That means that they will have to face the fears, guilt, hate, judgments and any negative thoughts that enter into their consciousness. They will have to face them, love them, and then release them to the universe. No one can survive this high energy area carrying negative feelings as excess baggage.

The particles of light in the band are very special, and they will actually help people by raising their frequency level and helping to prepare us for what we are going to face in the very near future. As the light particles enter our body and continue to move through it, they will help move negative thoughts and feelings, buried deep inside, out of our system. As they are removed and more and more particles enter our physical body, our frequency level will continue to be raised.

This will affect all living things, from the smallest microbe to planet Earth herself. The tiny organisms that are causing many problems and diseases will not be allowed to any longer. They will not be destroyed but will be transmuted into a higher substance or matter, as will all plants, animals and inhabitants on this planet.

Many of you have been well aware of the fact that the planet is going to go through her own cleansing. Well, the process will be the same for her. As the particles of light enter the Earth and exit the other side, they will remove all of the negative thoughts and feelings that have been buried in her body, the same as with the Earth's inhabitants. There will be countless trillions of particles entering the Earth to accomplish her cleansing. She is no different in that aspect from the human inhabitants that live on her. She will have to go through this cleansing process just as the other beings, large or small, or she will not survive.

The geological changes that will have already happened will begin the disposal process of the physical pollutions with the aid of the band as Earth enters the higher frequency area. This will have to be a slow process also. After the cleansing process is complete, we will then be closer to the One that we call God and be more able to be in His presence.

Changes in the Solar System

We have covered many things now during the last almost three years...about the past, the present and some future events. I know, Bobby, that you do not exactly understand the idea that they are actually all happening simultaneously. When you are finally brought into the fifth dimension, it will all become clear to you. You will be aware of things without having to experience them in the day and night cycle. The concept of time is a necessity to the ones living on the Earth plane in the third dimension. But that too will change. In your remembering, all the things you are curious about now will become totally clear.

The future, as seen by ones now living on the planet, is constantly changing because of the thoughts sent out from the consciousnesses of all beings on Earth. If you were in My position and could perceive all these thoughts and feelings, you would be amazed. Since everyone, including Me, are all connected by an invisible network of lines, all are aware of the thoughts and feelings of others. Nothing can be hidden from one another. That is the reason it is not desirable for ones in the third dimension to remember who they are – they would not be able to handle the results of knowing the thoughts of others. It would be utter chaos. The veil is there for everyone's protection, even yours.

Many who know this also know that the veil is thinning, which is why some are now beginning to pick up others' thoughts and why some are aware of what you would call future events. The veil will not be completely lifted for a while, and for many it will not happen at all in this cycle. It will happen when the Earth moves in closer to the center of the high energy belt. When the solar system enters the point I have called the center, it will remain in the highest energy of the band for hundreds of years. All physical bodies, not just beings on this planet, but all the planets, moons, asteroids and even the sun will be changed. Scientists are already seeing some of the results of the various planets moving into the higher energy levels of the belt. But mostly they are concerned with major changes taking place on the sun, which will become more pronounced as your solar system approaches the higher energy levels.

Some scientists are becoming frightened at the prospect of what is occurring on the surface of the sun as well as in the interior. Solar flares are becoming more frequent and reaching farther out into space. That is causing panic in those who are watching and recording these events. Some in the scientific community have determined that the sun has been a danger to the planet in the past and will be again, because they say solar phenomena happen in cycles. That is true, but they are not aware that these destructive solar events coincided with the solar system entering the high energy band. Even though the sun and its planets are now entering the band again, there will be no danger to the planets and their inhabitants that are completing their cleansing.

The sun has been important to the survival of all life on this planet, and it will continue to play a critical role in what is to happen to all beings, both animate and inanimate, as well as to the Earth itself. Nothing will escape the effects of the higher energy. The Earth and the beings that have stored up fear, anger, guilt and other negative emotions deep in their consciousnesses (mostly humans but also some animal life) will be the only ones that will have to go through the cleansing. All other life forms that have not harbored these emotions will only be affected in a positive manner. This includes the other planets, which are already being altered in unusual ways.

The sun's role, which will continue to be important as I have mentioned, will change during the coming event. He will be the receiver of much of the increased energy and will disburse it to the other bodies of the solar system. It will be similar to the role some beings have taken on as healers. They receive healing energy from Me, then focus that energy on ones they are working with so those ones can better heal themselves. That is what the sun is doing throughout the solar system. It is very important that ones that are ready for this energy prepare themselves to enter into a higher plane, in order to survive.

I have said that this energy will affect everything. That includes all manmade objects. Past eruptions from the sun have affected and even destroyed some satellites. Those eruptions that are soon to come into the atmosphere will be even more devastating. All electrical devices above the Earth and on the surface will cease to operate – appliances, electronic devices, computers, and especially automobiles and airplanes. That will take place as we approach 2012, which is closer than you think.

The sun will continue its pattern of increasing phenomena as your system enters into the high energy belt at an accelerating pace. Even though only a small amount of the population is aware of the higher level of energy they are experiencing, that will change. They already know that something is different. Most will feel an uneasiness come over them. It will not be pleasant for most because I do not see more than a small per cent of the world's inhabitants being prepared for what is coming.

As your system continues to move into the band I have been observing the changes in you personally, Bobby. You are not really aware on a day by day basis what is happening to you. But it is having more of an effect on you than you realize. Everyone eventually gets what they desire most. And you are well aware of your desires, in particular the one you have been focusing on. Since you have been keeping it before the universal consciousness, it will be made manifest before most of the planet's population knows what is going on.

The Belt's Place in the Universe

I know that you are not completely convinced about this not being the year that most feel it is – 2008. Actually they believe that because they have been told that for nearly two thousand years. During those early A.D. years no one was really sure what year it was. How do you determine it with a dating process that changes from what has been called B.C. for thousands of years, and then you stop and go the other way? Many were involved, but no one really knew the exact year. As I told you in another writing, it is actually now the year 2010.

I am going to continue to relate to you many more important facts about the past and what you have determined to be the future. Life is so much easier when you do not have to consider time – not enough time, running out of time, I need more time. When you think about it, there could never be a time when you do not have enough time. Time, like space, is endless. So how could you run out of time? Once you realize you are not on a schedule, you will soon be much freer.

Even though I have been telling you ones that the timing has to be right for these coming global events, that is not the same as counting time. There are just events that come together periodically in space, which is infinite. And these events are now in that position. Time has nothing to do with it. When those certain events do get together like they are about to do, then certain things will begin to happen. It is not like drawing a line and saying, "Okay, it's time." Things such as this happening, that has already begun, happen slowly. As soon as one event is past, then it is already on its way to another happening. It is a never-ending process.

When certain objects in space line up in a certain manner, certain things begin to happen. Astronomers have long been aware of this since the first civilizations existed on Earth. As you might expect, the knowledge was brought with them. When the creation process began, it was in My plans for just such a phenomenon. There is a lot more to it than most realize. Since everything is connected and everything has a consciousness – including celestial bodies – there is a great deal that could, and eventually will be known and understood by connecting with all these bodies in space.

As I said earlier in this writing, when certain alignments are made there is a different energy involved. The more bodies in the alignment, the more energy is involved. This not only means an alignment of stars or planets, but also clusters of stars and even clusters of galaxies. That is what is taking place at this point in space. I have mentioned this somewhat, but not at length. This is an alignment that has been in the process of coming together for millions of years. When something of this magnitude is happening, the alignment will continue for quite a long time. That is why I am saying that we will be in this high energy band for several hundred years.

While most all of the participants in this phenomenon are gigantic in size, some are of small configuration, and they will move aside more quickly. When that happens, the energy will shift somewhat. That is the reason that the energy is already changing – increasing. As a new player enters the game, the energy increases until we enter the area where the alignment is complete. Then it will reverse on the other side as we begin to move through and out. Then those players will fall away. Even if I do say so Myself, it will be an amazing experience!

Now you see why it will be an incredible period as the alignment begins to take place. The beings that do not participate in this event will have a long wait for the next turning that will come into alignment again. I will repeat: it will be a very long time before this alignment happens again. This alignment is for the ascendance of entire planetary systems and for whole civilizations to move into a completely different area where they can more easily evolve and move into higher planes.

This occurrence has never happened before. It is a first for this solar system and its inhabitants. As you explained in your first book, once this takes place, it will then spread out into other systems. As these other systems move into this alignment, they also will have the same opportunity to ascend as the inhabitants of planet Earth.

The one main difference is the fact that many of the other systems are much more advanced and have been preparing for this event in many cases for eons. They are more advanced in their science, as well as being more aware of their connection to the Source. They will evolve as you ones have, but since they are more aware, they will continue, for a time, to be more advanced in all things than you ones on Earth. You may wonder, if they are so aware of who they are, why this occurrence hasn't happened to them already. It is because although they have prepared for the alignment, it is not meant to happen at the same point in their evolution as it will happen for Earth.

This particular occurrence was planned soon after the creation process began. It was to be for this specific solar system at this exact time at this exact point in space. Most of the more advanced civilizations knew this; so they were able to plan to participate in this process. Some of you ones of this planet have been concerned with the overpopulation for a while now. Look how much it has increased in the last century – from something like 1.5 billion to about 6.5 billion at the present time. Do you think that has been an accident? Nothing is left to chance.

That has been allowed by the hierarchy of this system so that as many as possible could participate in this current event. Not only will this happen to the beings on the surface and under the surface, but also to the ones in thousands of spacecraft that are in the immediate area of your entire solar system. This is not only a possibility, it is an absolute certainty. This is what creation is all about. Once you move into a higher plane, then you can see the enormity of all this.

When I explained above that it would be a very long time before this event would return so that others could experience it, I did not mean there will be no opportunities for individual souls to advance to higher planes and continue with their evolutionary process. This process of evolution will continue until every soul I created from the beginning has returned to the oneness of creation – Me! No matter how long it takes. I have nowhere to go, and My patience is unlimited.

The Purpose of the Photon Belt

What I described to you just a few days ago about an alignment has something to do with the photon belt. You have referred to it as the Creator's high energy band to some extent in your book. At least one author has discussed it extensively in his book, and you received some information by another some years ago. Now you will receive the truth about this phenomenon straight from the horse's mouth (one of your phrases I like).

It is not a newly discovered observation; it has been known about for some time by some in your scientific community. But they only know that it is an area of high energy particles. To call it a photon "band" is mostly correct. It is a very large area of light particles that I created for a specific purpose: to raise the consciousness of all beings that I have created when they are prepared for the event.

It could be something right out of your science fiction TV series The Twilight Zone, but this band is for real. Any solar system that has at least a fairly evolved civilization will eventually be visited by this band. However, their system will not move into the area until a large percentage of the inhabitants are ready. They must be aware of the band's existence and the distinct possibility that they are about to enter into it.

Many of you ones on Earth have been made aware of this band by one means or another, although most have not accepted it as fact. But that is because its existence has been intentionally hidden from Earth's inhabitants so that they will not learn that it has to do with becoming aware of who they really are. Earth is a prime example of correct timing.

The photon band is not stationary in one area of the universe. You might say that it has a mind of its own, but its movement is controlled by the universal consciousness. So the path the photon band takes as it makes its way through the universe is not a haphazard one. Although most do not know it, when ones are completely in tune with the universal intelligence, they can move to any area of the universe almost immediately upon desiring it. It is the same with the photon band, which is actually more of a sphere. (However, the few who have known about it do not exactly define the shape; they only know that it is an area of high energy.)

The area of energy does not suddenly appear where there is a solar system; it moves into a preconceived location that the system is already moving towards. And as I have stated in previous writings, the closer the system gets to the band, the higher (more intense) the energy is.

This planetary system has entered into the fringes of the energy area before in its history, for a partial awakening. It was not ready to enter into the vastly more intense area in the past, but that is where this entire system, including the sun, is headed now. The photon band is so large that the whole solar system will be in the most intense area for several hundred years. What I am describing does not fit precisely with what others have written because they were only supposed to inform the ones of this planet of the band's existence. Now I am giving more complete information to you ones that want an accurate account.

There is much to tell about this high energy area, however, and I am not able to completely explain it in words that can be put on paper. The one that is writing these words has been given a more comprehensive picture. What is important to understand is the band has been designed specifically to raise your level of consciousness, and it will do that for all who enter into it. I have also said that individuals can and have raised their energy level without going into the band. But it is an extremely long journey, and only a few souls have taken that path to the higher planes.

Entire societies from far distant planets have evolved into incredibly high planes and are aware of all that exists in the universe. They have had that status in some cases for millions of years. But it also took millions of years for them to evolve to that level of spirituality. When I realized it was taking such a long time for created beings to reach that level, I decided that was not satisfactory and devised a faster way to move from one level to the next: the high energy area. And it truly is phenomenal. But the ones who take this path must be properly prepared to make the journey very quickly.

Partial descriptions of what to expect when you enter this area have been given to some. The ones who are not aware of the effects are having a rough time. That is not the way it was to be. The lack of information and the fact that Lucifer has kept so many of the population at a lower vibration are the reasons for so much chaos on the planet. Everyone must learn of the photon band's existence and how to make use of the energy. Time is of the essence. Because the knowledge has been hidden, there will first be chaos, as I just mentioned. But as beings become aware of the band and its purpose, they will adjust and be able to release many of the fears they have accumulated and just "go with the flow."

As I have said many times, I would have preferred that the Earth could have safely moved into the belt and completed this phase of her evolution without the cataclysmic events required for the cleansing. If things were to continue much longer without the correction Gaia must do, the end result would be much worse than the cleansing that is to occur. Due to what happens when objects enter into the band, at a certain point its effect is irreversible. If what is to enter into the band is not cleansed, it would eventually be transmuted back into the universal "melting pot" to be reused in some other form.

If that were allowed to happen to Earth, its absence from its present place in this solar system would be very difficult to overcome for quite a long time. The effects would be sent out in waves throughout the universe, and then the hierarchy would have to react to accommodate the rearrangement of much of this part of the universe. As some are aware, this situation has happened in the past with catastrophic results. That event took place in this very solar system. It was decided then that it would not be allowed to be repeated. Also, I have said that when it is time for a body to be recycled, it is planned well ahead of time so there can be plenty of advance warning and planning.

Three Days of Darkness

There has been much talk of the "three days of darkness" and the various possibilities pertaining to the event. The light particles are so densely packed together that they are impenetrable. That is why the photon band is so hard to observe. At a certain point as this system approaches the band, it will be faced with total darkness. As has been described by others, it will at first be like twilight and gradually become the darkest of dark. So inhabitants need to be prepared for that. But it will not happen suddenly. There will be ample warning and time for such things as airplanes landing safely. (Of course the results of the shift will determine if transportation is even available.)

Some believe that because the sun will not be able to penetrate the density, it will be extremely cold. But remember the density is made up of high energy light particles – photons. There will not be cold, only unimaginable blackness. After approximately three days of darkness, there will be another period of twilight. Then it will begin to be extraordinarily bright. Without some protection for your eyes there will be a temporary blindness. After a short period everyone will become adjusted to the light source.

From this point on everything will change. The sky, the stars, the sun and moon will be different. It will be an incredible feeling as your bodies begin to transform. The photons will permeate all physical matter, and everything will become semitransparent. Your senses will become far more acute. You will interact with all beings and begin to know each other's thoughts and feelings. This will be the most difficult adjustment. It will take time to get over the initial shock of being able to "read minds." But remember that everyone who survives to the point of moving to a higher consciousness will be prepared for this experience.

With a little bit of practice you will be able to create anything you want. But since everyone who survives will be a more highly evolved being, your needs and desires will not be the same as before. It will take time to build your body of light, and lots of adjustments. This will be a learning period that will last for hundreds of years. However, time will have been altered – you will not be aware of time and those hundreds of years will just fly by! And then a new adventure will begin. But that's another story.

9

The Role of Planet X (Creator)

The Planet, a.k.a. Nibiru

In your first book you were given information about the planet called Nibiru and its purported inhabitants. You were also told about the beings that had taken control of a large spacecraft linked to Nibiru. This craft has made periodic visits to Earth. Events pertaining to these two objects are indeed going to take place. They are closely related to the cleansing of the planet. This synchronicity is all part of the plan. (Actually there are more than two objects. The planet has satellites and that is where the associated spacecraft is based.)

As many have read, the orbit of Nibiru, or Planet X, is a very long ellipse that takes a considerable amount of time to complete. Since the orbit is only somewhat stable, its timing and proximity to the Earth can vary. This time around it will come close enough to cause the damage being predicted. In the past some of its passes were not close enough to do any significant damage.

Although its inhabitants know that certain things have happened with their planet's arrival in the past, they are only partly aware of the Earth changes about to happen. The craft I spoke of has a home base on Nibiru. The living conditions on the planet are very hostile, but the beings are able to survive there. It has no atmosphere and no water. The planet is larger than Earth, which is why it can cause so much destruction here without sustaining damage itself.

It slows as it makes its turn around the sun, and that is when the spacecraft leaves its home planet to visit Earth. It is responsible for the many legends of the past. They stayed long enough to be involved with the different civilizations of the time and brought the technology that made possible the construction of the amazing structures you find so mysterious today. Then they would return to their planet to await the next trip around the sun. They have had a substantial impact on the evolution of the Earth. There was never any interaction between those beings and other off-planet beings that might have been here at the same time.

Nibiru was placed in the galaxy eons ago to do just what it is about to do now. When I first noticed how much turmoil was being caused by Earth's inhabitants, I created a mechanism for rebalancing the planet when necessary: Planet X. I wanted to be able to control the amount of damage it might do, hence its slightly variable orbit. I once "manually" changed the orbit in the past because there would have been a fatal collision if I hadn't.

I have sped up the current orbit to avoid the almost total annihilation of Earth's population that would have occurred otherwise. There would have been few survivors on the planet; however, a large number would have been taken off and returned later to take advantage of the rendezvous with the consciousness belt. That part of the plan would not have changed.

So why did the part of the plan that involves Nibiru change? Ah, because of your concerns, Bobby, and your request on behalf of the inhabitants, although that may be hard to believe. You asked if there could be a reprieve of sorts and still accomplish the necessary cleansing without the planned level of destruction. I granted the request by changing the orbit of Planet X so that millions could survive on Earth and enter the consciousness belt without leaving the planet.

The ones that asked Me for My help when they knew Earth would be stripped of all but a miniscule amount of its population are the ones that would have been removed from the surface and returned later. As the time draws near, I will add more detail to this scenario, but you now know most of what to expect.

The Craft's Inhabitants

We will now talk more about the rogue planet, the large mass that is approaching Earth for one of its many rendezvous with it. I have already explained that I have altered its course somewhat to be more aligned with the amount of changes it must cause. I have also talked about the spacecraft that makes its home base there. It makes other journeys throughout this part of the galaxy, but it always returns to its base as the mass approaches Earth.

The last time it visited Earth, it did not make its usual extended stay. There had been a vast amount of change since the beings' prior visit – the time they had the biggest impact on the planet. They arrived here a long time before their home base arrived, and they were responsible for many physical changes in the lives of the inhabitants of the planet and for many of the structures that were far more advanced than the ones that were built by the locals.

The ones who make their home aboard the craft are due again before the passing of much time. This time they are mainly returning to see how much has changed since their last visit. Because they do not plan to spend any time on the surface, they plan to arrive only shortly before their base. They will be quite surprised to see what change has occurred. But the biggest surprise, of course, will be the huge armada of spacecraft that are here. They have encountered spacecraft here before, but usually only one or two.

I expect that they will just show up, find out what is going on, and soon depart. They will probably leave as their base planet passes by the Earth. In the past, if there was not too much damage caused by their planet and it was possible to help, they sometimes hung around to help the survivors. Otherwise, if the damage was too great, they would just leave. It only took them a couple of trips to figure out that their home base caused problems on Earth each time they arrived.

The ones on the craft are basically loners; so they do not interact much with other civilizations they come in contact with. They are mostly what you would call traders. They have a route they pretty well follow in their travels. Their main task is to maintain the upkeep of their craft. It is many thousands of years old, but they are able to keep it in relatively good condition through their trading. There are locations where they mine raw materials to swap for the finished materials they require, including food. Although it is sometimes hard for them to use updated parts for the craft, they have excellent means of adapting what they acquire to fit their needs. They have a first-rate workshop for that purpose.

What many will wonder about is how these beings have managed to live long enough to make all these journeys through space. First, they have the means for almost instantaneous travel from one location to another. As for living so long, they have a good understanding of the workings of the universe. They live for several hundred years, and when it is time to leave their worn out body, they go through the birthing process but with full knowledge and memory of who they are.

They are totally aware of Me and obey the laws of the universe. They are extremely intelligent, but more importantly they are wise in the ways of the universal consciousness. You may or may not have a personal visit with them. You will possibly only make contact with them to explain what is going on.

Gaia's Involvement with Planet X

Gaia and I have said previously that only Gaia was to be involved in the planetary changes. But I was just not ready before now to reveal how Gaia and Planet X would relate to each other. But they are indeed related. If it were not the case that Gaia could act as a sort of moderator, the results of the interaction would be chaotic. There would be no rhyme or reason to how things would end up. Actually, how things turn out will result from the three of us, because I have already guided Planet X to the correct distance from Earth to accomplish the results I desire, with Gaia's help.

Normally, I would not say that one's part was more important than another's. But long ago I gave Gaia the authority over the proper placement of energies to obtain the desired results. That energy will be a combination of energies from Gaia, Myself and Planet X itself. Since I have complete faith in Gaia's ability to accomplish what will take place, I will be acting mostly as an observer. Then I can be of assistance if she asks for any kind of help. I have already seen the final results, and as expected, these conditions will accomplish what I desired.

The Transformation of a Planet

100

10

Enter Lucifer

Who is Lucifer? (Creator)

Lucifer is one of the main players in the game, and has been since he was first created. Now there will be many that will be angry that their existence has been called a game. (Mostly those will be the ones that will be playing their game over and over.) It could be called many things, but by calling it a game, which to Me is very appropriate, it makes the situation a more joyful experience.

Other than Myself, Lucifer has probably been cussed, discussed, and talked about more than any other of My creations. Lucifer is not only known by that name, but by many other names such as the Evil One, Prince of Darkness, and the negative side of creation. But before his so-called fall from grace, he was known in the beginning as the brightest of all angels – an archangel even – by Christians.

Yes, he is one of the archangels and was created as one of the "second cause", as you have termed it, along with you and Sananda and the rest of the archangels. When he agreed to implement his part of My plan, he would then be thought of as a fallen angel. I agreed that some of the archangels could go with him. Not much is known about these others, but their part will eventually be made known. This has been at their request. That is why there are only seven archangels known by the ones of this planet.

The time will come when the true facts about Lucifer will be made clear to all ones of the planet and beyond. There will be much said about the part that you and Lucifer and the rest of the team played in this game and why the game was played. Even though you are not able to feel it now, there is much love between you and Sananda and Lucifer. It will be very heartwarming for the three of you after this is over. Even now there is no resentment. Well, not much!

There are some things that happened between you and Lucifer that you are not really aware of. I will now take you back to the early days of the creation process when Lucifer was more or less just going through the motions of his part in the creation process. At the time when you and the other archangels were created, you were just beings that I thought of as only a part of Me, not beings separate from Me. You were just "out there" in a place I could observe differently from before you were released from Me. So what you were doing in the creation process is also what I was doing. You were the part of Me that was experimenting with different ideas. There was no thought of being separate from Me. The idea of free will had not occurred to Me.

As time passed I began to feel a certain sense of separation, especially from the one that was eventually known as Lucifer. Lucifer did not have thoughts of rebellion or thoughts of taking control at that time; he was mostly concerned with himself. He preferred to be alone with his thoughts and that was all right. At some point you, as well as Sananda, realized that there were differences between Lucifer and the other archangels. You tried to tune into his thoughts with the intention of communicating with him. For a long time he just ignored you, but you were very persistent.

After a long time of trying to have Lucifer acknowledge you, he finally did respond...but not as you expected. He politely told you to mind your own business! No being had ever received that kind of response from

another. That was your first run-in with Lucifer. The next time he told you to stay out of his mind he was not so polite. With much stronger feeling, he made it clear that you were not to try to make any contact with him ever again. From that time on he was a recluse.

After Lucifer arrived on the planet Earth many thousands of years ago, he did some planning, but mostly he just waited and took advantage of certain things and events that happened. (You have already written about the results of those events. They are basically the way things happened, with some minor differences.) You and Lucifer agreed that you two would just stay out of each other's way until it was necessary for the two of you to have your "showdown." (We'll come back to this later.) Lucifer played his part to the hilt. He helped the ones of the planet that he was involved with create all of the places and events that most now see as real. They have created "hell", Satan, the various devils and many other little things that could be used by the Church to control the population.

Lucifer Playing God (Creator)

Let's briefly go back to the early life of Abraham. He was one of many wandering sheepherders. He was the unspoken leader of several small groups trying to eke out a living in a desolate area. But there were a few very green and fertile regions that these groups were unaware of. The Pleiadians decided to give Abraham some help. Their motive was not completely humanitarian, but they made promises to Abraham and lived up to them. They were there for many years; so Abraham became quite wealthy in terms of possessions. He controlled much land with a large amount of livestock.

After the Pleiadians left the planet for the last time, I decided that I would make Myself known to Abraham because it was time for a spiritual rebirth on Earth. Before Abraham there was basically no spirituality anywhere except for several small pockets of aboriginals, who only had a knowledge and memory of Me. So I slowly made Myself known. It was relatively easy because of the presence of the Pleiadians.

For many years I guided Abraham and after his death, his sons and grandchildren. After a time, as they moved away, some split off, formed a new religion and began to drift away from their spiritual roots. I felt it was time to leave Earth. The birth of the Hebrew Nation was well on its way with all of its good times and bad times. Why do you suppose this nation that was brought forth by Abraham, who was a good, loving being and loved by all, had so many bad times over the next several hundred years?

At the time of My almost physical sojourn on this planet, Lucifer had been here for several thousand years. He had been gearing up to do what we had agreed upon long before. When I said I was leaving the planet, that was only one of the subordinate selves that I had created for tasks like guiding Abraham. My consciousness, of course, is everywhere at all times. So I knew about Lucifer during this time, but I never allowed our paths to cross. I could see his plan forming and knew what he would do as soon as My presence was gone. He had decided to take My place. He had been planning it for some time and was very aware of what I had been doing. He had never played god before; so he paid attention to My activities so that he could move right in.

Although he did not know when I would leave, he correctly determined I would leave sooner or later. It was a very smooth transition – he would play his part to a T. This was the opportunity he had been looking for. He knew he had plenty of time; so he did not rush into what he had been planning. He planted seeds of thought to slowly transition from My type of being God to his type of playing God. Once he felt he was ingrained in the minds of the ones of the Hebrew Nation and had his hook set, he started giving it a yank. Little yanks at first, but once he knew that he had their attention he pursued his plan.

When I was working with Abraham and his generation, they wanted to know what to call Me. I replied that I had no name and did not care if they called Me by an Earth-created name or not. I told them I was their Creator and the Creator of All That Is, has been and will be. I just wanted them to recognize that fact and that they were part of Me. I told them I existed inside of each and every one. I required no temples, no bowing, no worship. I only wanted to be acknowledged for who I was. They finally realized that and began to live good lives, loving Me and all they came in contact with. I guided Abraham and had many conversations with him. When he passed on, things began to change, and I soon left.

Things would be different with Lucifer. He wanted to be praised, exalted and eventually worshipped. He decided he would need a different name. It would not be good to have his new subjects call him Lucifer because he was already known by that name among certain ones on the planet who could ruin his plans. Regardless of the fact that some say the name Jehovah as a name for God did not exist until many centuries later, that is what Lucifer chose to be called. Of course, he was known as Yahweh to the Hebrew Nation.

Most people know the results of Lucifer's experience as the God of that nation. He planted lots of seeds of discontent that grew down through the ages to the present. As I said earlier, this could get a little nasty for you and Me. This will not sit well with a big percent of the world's population, especially with the Jewish nation.

Lucifer and Jesus (Sananda and Creator)

Creator:
   Not much has been written about Lucifer during the time when Sananda was on the surface of this planet as Jesus. Lucifer was not very aware of all My plans for

Earth. He was made conscious of his part, but that had been far back in the early days of creation. Because he was mostly focused on his own plans at the time, what is in store for Earth was shoved to the back of his memory. But once Jesus was born and Lucifer saw what was taking place, he began to remember the plan for Earth that I had revealed to him. He also remembered that Jesus had been prophesied to come here as the savior of mankind.

Lucifer knew Sananda was an archangel, but the two of them had never had any encounters to speak of. So it took him a while to realize that Sananda and Jesus were one and the same. Needless to say, Lucifer was rather curious and a little surprised when Sananda came to this planet through the birthing process. Once Lucifer figured it all out, he knew he would have to pay close attention as Jesus grew into manhood. He could see that Jesus' role would be important to what he was doing on the planet. As the Bible relates, Jesus was to do battle with Satan (really Lucifer). They did have a brief confrontation, but Lucifer preferred to back off and watch Jesus from afar.

After Jesus was crucified, buried and ascended into a higher plane – where he remains – Lucifer was sure that Sananda (as he is now known) was part of a major event in the future. He had known for a long time that you (Michael) were here, and by observing the two of you he was able to figure out a little of what was going on. However, as he became aware that you were part of a whole team – the team I had put together so long ago – he got a bit alarmed and tried to contact Me. But I was not ready to communicate with him yet.

When Lucifer began to pay attention to the team, it didn't take him long to sort things out. He realized that the major event had a lot to do with him and what he was doing. Until that time he was certain that he would succeed in winning over all the souls on the planet. But a little doubt crept in, and since Jesus' death Lucifer's efforts have been frantic even though his labors have paid off. His success has made him more confident than ever about his goal. You see the results of his increased effort everywhere. He recognizes there is little time left for harvesting souls, but he will not slack off until the very end, and he will not give up easily.

Sananda:

Much has been written about Lucifer and your association with him. And to a lesser degree with me, even though I have had more associations with him than has been generally known. The three of us have spent much more time together than you have written about. After all, fifteen billion years is a long time.

You and I are aware of the fact that we are actors in this play that is in its final moments before the curtain falls for the last time in this particular play. Lucifer is only partially aware of this, but he is so involved in what he is doing that he does not want to believe it and is keeping the truth hidden from himself. I will briefly discuss the fact of your "showdown" with Lucifer, which has to take place. In the beginning it will not be a pleasant affair, as you have been told and are aware of. Enough said.

After Lucifer makes it right with his Creator (I know, there is no right or wrong, but most beings still need to have a choice), and even more important makes things right with himself as far as forgiveness, he will then be a part of our future. In Earth time it would be a considerable length of time before he has "paid his debt", but we will soon be leaving the concept of time behind us.

So in reality it will be quite soon when the three of us are together again to go off on another adventure. I was not to discuss this until Creator knew that you were about to have a remembrance of it. I only jolted your memory a little. So we can now discuss this matter because it is time for all to know. Once enough of these kinds of remembering take place then they will begin to happen more frequently, until – Wow! – it will happen all at once – full remembering.

Lucifer's Last Chance (Creator)

The ones in control will be given every tool in Lucifer's bag of tricks – anything he can think of to postpone what he knows to be inevitable. He feels that the more he can delay, the better his chances of winning the battle. Although he is feeling pretty good about himself and what he has accomplished, he is still plenty angry about our destroying the shield that he was sure would withstand whatever force we might use against him. But he has adjusted, and with his constant bombardment of negative energy to the inhabitants, he feels that he is gaining.

I have not mentioned this before; so it is now time for Me to explain this to you. You know that whatever you desire most you will eventually receive, if it is sent to the universe with passion and desire. Since the universe only recognizes this desire, it does not see the difference between what you know on the planet as good and evil. Lucifer has only one desire, as you have only one desire. His desire is to have complete control over all the subjects on Earth, that he already sees as his. He knows how to play the game very well.

Therein lies what this is all about: whether Lucifer can use the power of the universe to influence all souls and continue until they are completely under his control. To do that he will have to extinguish the Light of My Being that shines in each and every one of My created beings. His one great advantage is that he does not recognize the universal law of free will, even though he will try to use it against us. Now you see what you are up against.

But I also have an advantage, one that he chooses not to use: I already know the outcome of this conflict. I can see where he is not looking: into the future, as you ones term it. He is focused on what he is doing now and not on the end result. You have seen the end result, just as I have. Lucifer cannot see it because he is not looking. He is concentrating on keeping his control over the thoughts of the ones that accept it.

We will not do battle with him. That would only prolong things. That would be asking the universe to continue the battle that has been raging for eons. It is time for it to end. And since you have already seen the end result, as I have, then you already have the results – the gift of the universe of bringing it to an end. You could say that the universe is opening wide its gates for this event. Are you now beginning to see what the adversary is up against? These are not afterthoughts; this has been the plan from the start.

I want you to be fully prepared because I have allowed the adversary to be the same. I want them to give it their best shot. I want them to know that I did not put any limits on them other than the one I set in the beginning: Lucifer could not directly take any human's life. He could, however, influence others to do whatever he was able to convince them to do, and to obey his commands if he was able to take control of their lives. And he has succeeded in doing that. You and the team actually have more limits than Lucifer, but I am giving you all of this extra help and making you aware of it. You will need it.

Monitoring Lucifer (St. Germain and Bobby)

February session

St. Germain:

I spend most of my time monitoring Lucifer's daily affairs. He has been aware of this almost from the beginning. At first he tried to ignore me, as he did with you in some of your encounters in the distant past. But lately his behavior has changed. He seems to be overly eager for me to keep tabs on him, sometimes appearing to be putting on a show for me. It is just not like him to draw attention to himself.

I have discussed this at length with the other team members, and they are at a loss as to what Lucifer might be up to. We know that anything, and everything for that matter, is always to his advantage right now. We thought you might have something to suggest about this new development. As far as I can tell, he has not slacked off on his main focus of commandeering as many souls as possible in the short time he has left. We feel he has something up his sleeve, so to speak.

Bobby:

You probably have already thought of this possibility, but this is what comes to mind off the top of my head until I think about it for a while. As we enter farther into the high-energy/photon belt, it raises our consciousness level and increases our awareness. Could this be having an effect on Lucifer that we have not considered? I'm not saying his consciousness is being raised, but there may be a response unknown to us. Do you know whether he has ever previously entered into the belt at one time or another?

St. Germain:

That is something we had not contemplated at all. We certainly need to look at that and see what comes up. With input from the new beings that are here now, we just might be able to come up with something. I will check the records and see if Lucifer may have entered into the belt in the past. We will get back to you on that.

Other than his new acceptance of my keeping tabs on him constantly, he is just doing his thing. And doing it non-stop. Lucifer has demanded that the universe fulfill his desire to have complete control of the planet. Although you were aware of that, it was a surprise to the rest of us. He continues to send out his negative energy, but he is still waiting for a response to his demand from universal consciousness.

Bobby:

What if he is trying to change his image in a way that might affect his demands on the universal consciousness?

St. Germain:

Yes, we did discuss that; it is a definite possibility.

<u>March Session</u>

St. Germain:
I am still keeping an eye on Lucifer, and he continues to amaze me. He is still reacting as strangely to my presence as he was the last time you and I talked. Even though I am the only one keeping an eye on him, all of the others on the team are curious; so I keep them informed, and we have been discussing the way he is acting.

We have talked a lot about the fact that Lucifer might be affected in an unforeseen manner as we move closer to the most intense part of the consciousness belt. We all feel that is true. We could not come up with a viable reason that he would not be affected. After all, we have been told by our Creator, the creator of the belt, that no beings will escape its effects.

So now we are wondering exactly how the beam will affect Lucifer, as well as the others of the dark forces. Since it can or will affect everyone differently, at this time we can only make an educated guess as to what it will do to Lucifer. We have not brought it up with Creator, although we have considered it. We expect He will give us an answer through you when He is ready to. If not, I will just keep on observing Lucifer's thoughts and actions.

As to his presence with the elite corps, we have also considered that Lucifer might just be trying to throw us off regardless of his new personality – to keep us off track as to what his real agenda is. He is paying close attention to the elite corps. He does not do any of the actual planning in situations like this. He just gives them ideas, then feeds negative energy to them to keep the ideas moving in the Elite's consciousness until they take shape.

A particular plan has been on the burner, so to speak, for quite some time; so it is a mature, well thought out plan, and we expect it to succeed. Since Creator has said it is an event that will happen, we assume that He has already looked at it and seen its success pattern. (Creator has just interrupted me to say that the event is successful as far as the Elite are concerned. That rather settles that.) This may not be the event that will trigger our appearance; it may only lead to it. So we have to be very observant now, and be at the ready at all times. I expect that our plans will quickly be put into focus.

11

Michael and Lucifer

"The Showdown" (Creator)

There are facts that you are beginning to remember about Lucifer and his place in all of what has happened and is about to happen. Actually, it has been happening for many centuries, but it is now about to come to a head. Lucifer's part in all that has taken place over the last several million years was part of the ongoing plan, or *plans* (because I have made changes over the millennia).

You could say that he is one of the members of My team. He agreed to do what I laid out before him, but I gave him free will to do it mostly his own way, to do "his own thing." Now, I did have to step in as you have written about. For a limited time Lucifer did have some grandiose plans about himself when he thought he could pretend to be Me. You were aware of what I asked him to do from the very beginning. You became angry with him about what you could see he was planning. That was the confrontation that the two of you had. You took him too seriously. He really knew that he would not be capable of taking over My job. That's when you decided just to erase that from your memory.

You were aware of all of this and also what I am about to tell you, but you decided that you did not want to remember any of this until the time was right. You felt that you had enough to do, and to be concerned with, without that to think about. I agreed with you and allowed that to happen. But now the time is right to remember.

For a time the game with Lucifer and his "evil" ways will continue. But I do not foresee any trouble with him. You have been told that there will be great danger in what you will be doing. Now, there will be a confrontation between you two, but that is not where the danger comes from. The danger will be from ones on the planet – the dark forces. We have talked a little about what will happen with Lucifer and his legions. But it is mostly between Lucifer and Myself as to what will happen to him. He was aware of this when he agreed to his part in My plan.

You are beginning to have thoughts about your showdown with Lucifer. I told you it would be different from what you might have expected. By the time he reaches that point in space with all the others, he will have choices. He can ride it out and see what happens; he can have a confrontation with you and then later with Me; or he can listen to what I have to say to him through you. You have already reminded him of his part in the plan and how he kind of got carried away with himself. You will then explain to him what is happening and give him more choices. These will be the most serious moments in his existence.

Since there will no longer be any separation, no good or evil, Lucifer has two choices at this time. If he chooses to remain as he is, he will go with the others to the waiting area, but his waiting period will more than likely be longer than most. At some point he would then be recycled back into the universal consciousness and return with no memory of who he was. He will begin a new existence as a new soul. Or he can come back to Me after his own judgment of himself and be welcomed back with all the love that I possess. There would be a short period of cleansing, and then he will join you ones if he so desires. Then the three of you will be together again.

So there you have it, Bobby, your "showdown" with Lucifer. I will now say that all will have the same choice as Lucifer to come to Me. The biggest danger for you is the ones in the physical that are not willing to know what their choices are. They will not easily give up their dream of total power. They have many sophisticated weapons available, so they will not go down easy. Just be aware of that. Of course, we also have our own means.

Lucifer has been trying hard to break into your defenses, with no luck, to find out exactly what is going on. He thought he had succeeded at one point when I allowed him to approach you. Now he has redoubled his attempts to penetrate your shield. That is why it is so important for you to stay on a level above him. I am helping you with this whenever we are communicating. He has a strong feeling that there will be a confrontation between the two of you. He had not paid much heed to the feeling in the past. But I have not told you this before: at that time I will bring him into the presence of all of you and then later make Myself known. Only My presence will cause him to admit defeat. I will not tell the surprise ending.

Lucifer and the Photon Belt (Creator)

I have been watching and also adding more protection now that the main event is about to occur. Lucifer is aware of this and is making these last few days as profitable as possible. You have been wondering about the effects that the photon belt might have on Lucifer. He has been kept so busy doing what he came here to do that he has not been aware enough of the possibility that something might affect what he is doing. He has been doubling up on his efforts to achieve his goal. He knows that something is changing. His followers also know something is changing, and some do know a little about the photon belt.

But you and the other team members were right when you assumed that the energy would affect him. There will be a certain point in space that we will reach as we continue into the depths of the beam. Then everyone will be aware of its effects. Lucifer will know, but he will be able to fight it off for a while. Eventually, it will become too strong for even him. That is when he will come to us to find out what is happening. Then you and he will begin to square off with one another.

What will happen to him when we reach the "point of no return"? The consciousness beam will only affect the physical in what you call death. It will affect the spirit or the soul in an entirely different way. That has not been discussed before. I have said that the ones who will not awaken from the density of Lucifer's control will be placed in an area likened to suspended animation. They will not be aware of anything until the time that I will release that suspension. Then they will begin a completely new existence.

This is where Lucifer will remain if he does not come to a remembering, mend his ways, and raise his consciousness level like everyone else. Now, the ones that did not drop down into the density too far and can see their mistakes will be the ones that will be moved to other planets to continue their evolution. Some of this we have talked about, but I am just telling it a little different way.

Bobby's Conversation with Lucifer

The author:

For some reason that I could not quite figure out, the last one or two writings that I had received just didn't seem quite in order. Not that the information was wrong; something just seemed different. The feeling wouldn't let go. Then Sananda sort of "popped in" and started talking (at least I thought it was Sananda, but coming to my attention a different way).

He told me that he and the other team members had gone over all of the facts about the event planned by the elite group that was to take place very soon. Their conclusion was that the event was not going to happen after all. He wanted to know what I felt about it. I hesitated to answer right away, but I had had doubts, even though Creator had said it was almost certain to happen.

I was beginning to get the familiar feeling that something was not right. It just did not seem like Sananda. Then all of a sudden there was a dramatic change in his voice (energy). It was a very deep, almost guttural sound. When he called my name I recognized the voice as Lucifer's. Of course, my first thought was, "How did he get through the defenses that completely engulf Taliesen?" I didn't panic but I was concerned. He was allowed to penetrate for a couple of minutes a few months ago, but I was still surprised. He then said that he was permitted to enter into a limited part of my space to talk with me about a few things.

We had not had a conversation or an exchange of thoughts for a very long time. But he wanted to tell me about his part in our Creator's plan. I said, "Let's go all the way back to the beginning and start there." This then is his story about himself and his existence for all those billions of years. Creator had told me some of the facts about the very beginning of creation, about how Lucifer felt he was not suited to being a co-creator. Nevertheless, he did what he was created to do for a long time.

Lucifer:

I never liked the idea of being a creator, even though I did as was expected of me. It just felt that either I was different, or I wanted to be different. I say "different", but I did not really know what "different" was. My thoughts were not about creation; they were about something other than what was expected of me. I just did not fit in. This went on for a long time. I was aware of my Creator, but it was just an awareness at first, nothing else. I kept wondering, "Why isn't there something else?"

As time passed and creation began to take many diverse paths, I started to feel that there *was* something else for me. Eventually Creator came to me and explained the plan that He had been mulling over at length. This was after all the created beings experienced the sense of separation. After the Creator had outlined what would take place in the distant future on an obscure planet, He told me what He desired of me. I accepted without even a pause because I sensed the "difference" I had been looking for. For one thing, we had never had an exchange of thoughts like this before in all my billions of years.

While He told me what I was to do – my part in His plan – I was so involved with what I was feeling that I didn't really pay close attention to the details of the plan. I was supposed to "disappear" for a time, to become unknown to the rest of the universe. That was exciting to me. I was truly unique, an exception among the archangels. I was brilliant and I knew it. I also let every other energy I came into contact with know it. I could now do what I pleased; I could be alone with my thoughts – no creating.

During the time that I was gone from the rest of creation, I thought about what Creator had said - not so much about His plan as about how it would affect me. I knew that He had created another self so He could be closer to all of His creation. Now I knew that we were all connected as one; so He could not actually be closer, just different. He was everything but also different.

There were at this time two polar energies – the positive and the negative; the universe had been created this way. Creator had said His plan was that there would be two ways that life would evolve, two paths that creation could follow. The positive path was the light of creation – His way. He wanted me to represent the other path, the negative direction in His plan. This would eventually be called the way of darkness or the dark side. My thoughts led me to understand that this role made me very different indeed.

Then the Creator came to me again and told me it was time for my "entrance." His created beings, at least the ones I had been asked to observe, had a new awareness. Over a considerable amount of time these beings learned they had been given free will, and their "separation" from the Creator was almost complete. During this time they had created a highly evolved civilization with lots of "toys." They had developed space travel and discovered that there were many other advanced civilizations. They began to exchange technologies with one another and to engage in extensive trade. There was no violence in the early days, but it was about to begin.

After thousands of years of observation I became aware of rumblings, turmoil. As trade expanded, many species became involved. Although there were differences in physical appearance, they were not so important after a while. The loneliness of space travel leads to similarities being more important. Interaction between the species resulted in many varieties of mutants. Since they were not welcomed by either of the original species, space became their home. I could feel the tension, fear and hatred that some began to show. This new type of energy was stimulating to me. I felt I was gaining power from it; so I stayed closer to these ones.

I soon realized I could increase the negative energy by sending my thoughts to them. As the energy increased, I grew more powerful. They took to fighting among themselves until in anger it finally happened. One of the mutant's life force was accidentally terminated by another – the first killing. At that moment I felt a tremendous surge of energy. Wow! I urged others to do the same as my power got even stronger. They started killing each other for no reason at all.

Then someone realized that if they continued they would totally destroy themselves. So they joined forces, became the first pirates, and eventually triggered the first space wars. Originally there were only primitive weapons available for killing because despite the amount of technology they had then, they had never thought of building weapons. But a new industry developed and massive weapons systems were devised. After a time weapons were designed that were capable of destroying a planet.

During all this time, I felt the Creator's presence, but He did not interfere. I took His silence as permission to continue what I was doing. I was the "other path" of creation, the opposite of creation. I was a destroyer. I believed I was doing what I was destined to do. I was good at it, and I was enjoying myself. The idea that I was doing something wrong never entered my mind. I felt that I was now equal with the One you refer to as God. But things were about to change.

For the first time I became aware that some beings were conscious of my presence. That was a big surprise. I demanded to know who they were and learned they were the Reptilian race. We had brief encounters over time and eventually came to an agreement of sorts – an agreement that neither side expected to keep.

When the "star wars" became really nasty and there was a chance that a planet could be destroyed, the Creator stepped in. He gave me a very firm "Stop! You have gone too far." I complained and insisted, "You cannot interfere with my free will." He replied, "I can if I am asked, and thousands have asked." That was when you began meddling in my affairs again. It had been millions of years since we had last been in contact with each other. You came to me and asked for my help. Now that was a laugh! You really expected me to help you.

Then Creator gave me an ultimatum. We haggled back and forth, and the result was that I ended up here, in control of this planet...with limitations. As with the last time He sent me somewhere, I was to observe for a while. I could do almost anything except take the life of a physical being. I could plant thoughts in the minds of all beings on the planet, including the suggestion to kill. As you know I have been quite successful at that. I am here to win souls over "to the dark side" – as many as possible in the time allotted. And I know that time has just about run out. But I am determined to win; this planet will be mine!

Bobby:

Do you remember back when Creator asked you to be part of His plan for this planet? His plan was to let the inhabitants sink to the deepest level of the third dimension and then come back to the light. You were to try to win over as many souls as you could to prevent the return to the light. He gave you this one last chance to see how much power you have.

Lucifer:

I vaguely remember that. After I observed for a while and He gave me the go ahead, I wasted no time in my attempts. I soon found out you were here and that we would have a showdown, like "Get out of town by sundown." I am aware of the old westerns, as they are called. I loved them as much as I can love. I control the most powerful on the planet, and I helped them instigate a lot of wars. The part I liked best was being the so-called god Jehovah. When Creator left the physical planet, I saw the chance to become the god of the Hebrew nations. I really had them going for a while.

When Sananda made an appearance here and I recognized that he was Jesus, we had a few dealings. But when I learned you were here too, I knew something big was going on. I went back to gathering souls with a concerted effort. As you see, I have had considerable success, even if you did manage to destroy my defense shield. I spent many years building it and thought it was impenetrable from all outside sources. I never imagined it would be destroyed from within. But I will still succeed, by furnishing energy to the ones I control on Earth who are doing my bidding.

Creator asked me to relate this part of my existence to you now and opened a space inside your "fortress" for a limited time. It appears that we are to have that showdown soon – a "winner take all" event.

12

The Plans of the Elite

Who Are the Elite? (Creator)

Why has the Elite been planning the subtle takeover of the planet for so long? Basically they are just covering all the bases. They are preparing to come out on top no matter which scenario unfolds. So there are members of the elite corps in the scientific and psychic communities to keep abreast of all possibilities. They are intertwined with all the affairs of Earth more than you might imagine. What the Elite do not own outright, they can control by means of their tremendous wealth. They just "buy" the individuals in key positions in government, academia, science, industry and religion. The Elite are among the first to know of any new discovery in these areas. And they control how much of that information, or intentional disinformation, gets to the public.

They are not only aware of extraterrestrials; they have made deals with the ones that are here that do not have the best interest of the human population at heart. These off-planet beings do not have a mind to take over the planet. They are interested in certain research and the elite corps has made that possible. The Elite have cooperated and have received much in return, including warnings about things that are to happen to Earth in the near future, helpful ideas and off-planet technology.

However, the ET's are not completely trustworthy; so several events are being planned for. The Elite are behind the construction of underground facilities all around the globe. They are helping the governments finance them so that thousands of people can be housed in the face of a major disaster. It wouldn't do any good for the Elite to survive if no one is left to control or manipulate into performing the menial tasks.

Even though the Elite believe some form of cataclysm is about to happen, they are making other plans in the event it does not happen. That is the manufactured scenario, or set of scenarios, that will hasten their takeover: the failure of the US dollar and the skyrocketing price of fuel and food. They want to control those items first, and then they will slowly instigate the subsequent plans. They are in the final phase of tying up loose ends on all these projects.

Lucifer will play a part in all of this. The elite corps are the main players doing his bidding, some knowingly and some not, as I have said. However, he is not being honest with them either. So much deceit, so many lies. It is no wonder the Earth is falling deeper and deeper into the density.

The elite group's agenda is in place. As they tackle it in earnest, there will be plenty of ones that will be receptive to whatever the Elite have to offer in the way of money and power. The timetable is set for an event planned that will get the attention of the planet's entire population. When it occurs, it will be time for you and the rest of My team to make your presence known.

More About the Elite from Sananda

On certain physical levels there are meetings now taking place whose results will have devastating effects for most of the inhabitants of the planet. These meetings are the result of many years of planning by the so-called elite of the planet – the ones that have slowly taken control of most of the affairs of business and government: the large conglomerates, multinational corporations, and so on. Most ones in government are not aware of or part of this group. They are only "wannabe" figureheads, so to speak. Mostly they are under the control of the power group.

This group has been called many names, but they do not go by any of those. They are just the ones in control and directed by the "king of the dark ones." Some are aware of this, but most are not. Many have made agreements with the dark forces many years ago, in return for the riches and pleasures of this life. They are not concerned with their souls beyond the physical. They will have a price to pay.

A few families pretty well control what is going on at any one time, especially in the finance world. They can control most of the world's financial dealings that take place every day with phone calls, and nobody gives them any backtalk. We are talking about huge amounts of money – money that moves around the world every day. Many institutions just need to borrow money for a very short time; so trillions of dollars (and of course other countries' currencies) is loaned every day. When you talk about that much money, even a very small percent of interest being made can earn millions of dollars daily. The amazing thing is that most of the world's currency is electronic monies that are in the airwaves a big part of the time; so the actual money is not transferred. So how does anyone know whether that money even exists? Very interesting, wouldn't you say?

This group has family members in most of the countries' governments, industry, and even in the largest of the world religions. They have worked for many years to get as many of their people as possible into positions of power all over the globe. Where they do not control the government or organization outright, they have managed to buy those who are in high places with offers of great wealth and power. Much of the wealth has been distributed; the promises of power will be fulfilled, or so they say. When these "bought" ones are no longer useful, they will be disposed of. That is the way of these "evil" ones, as they have been called. So the table is set, the cards have been stacked, and the game is just about to begin.

Mostly, Bobby, I want to assure you that what is now taking place, as far as our Creator and we of the team are concerned, is what has been planned by the ones who are under the command of the dark forces. As I explained to you a few days ago, they were having meeting after meeting and are almost in a panic. As Sanat explained to you last night, we are monitoring them and will start informing you of their plans and the contents of the meetings. The results of the meetings will be bad news for the majority of the inhabitants of this planet, since their plan is to start implementing certain events now very quickly. As you know, we will not directly interfere with their plans.

What it will bring about is the landing of our craft, which will end all doubts about the so-called extraterrestrial beings. This we have discussed many times: a certain event, or events, will result in the decision to make ourselves known. We know this will cause mixed reactions among the general public and anger many of the heads of government, as well as some leaders of the Church.

As I have said many times, all will soon become clear and you will have a complete understanding of what is now taking place. Even though this is our project, our Creator will be advising us and suggesting certain things. After all, it is His plan. But it is up to the ones of His team to implement His perfect plan. Don't be discouraged; be happy. Continue doing what you are doing, but be ready to move at a moment's notice.

The Preliminary Plans (Sananda)

So much is taking place around the globe that it is difficult to keep up just with events that will be affecting us in the near future. Many of the groups that are aware of the upcoming changes are trying to figure out ways to

take advantage of them to fit their agenda. It seems that just about all have pretty much the same motive, which is "How can I figure a way for it to benefit me?" instead of "How can I be of service to others during the coming months as things that have been predicted begin to unfold?" Lucifer is having quite a lark as he feeds all these egos with the desire to only think of themselves.

It is a little different with the main players that have been waiting for this event for centuries. Their objective is to serve their master, and they have been doing that for all those centuries. Also, we have mentioned these ones that are all part of the same group but there is much infighting. They have finally come to the conclusion that they need to be in agreement instead of arguing among themselves about who is number one.

Of course, Lucifer thrives on this dissention, but now the families have come to terms and formed the board to work out their disagreements. They have accepted the fact that one person cannot be in charge and give all the orders. They finally awoke to the fact that it was all an ego thing. And now that time is running out, they had to come to an agreement on the issues and put the agreement into action. Acting on the agreements is what's taking place at this time.

That brings us to the meetings that we have informed you about. In the beginning the members of the group were just feeling each other out. No one wanted to commit at first even though they all knew the reason to be there. It was their livelihood. The military consisted mainly of high-ranking officers. But there were several low-ranking officers who had been brought up to date on the true facts, and they were entirely accepting of what was going on behind the scenes.

As has been said, the war machines that have been going on mostly at full blast for so many years will not be allowed to come to a grinding halt. There was some panic when the cold war suddenly ended, and the present situation is similar. If the war machine shuts down, many would be out of a job, and that is all they know. Some would be able to go on speaking circuits, but only a few. As wars die out, the groups that are fascinated by wars are okay with just the planning and talk of wars. The problem is that such a large percentage of the civilized population is fascinated by war.

Then the owners of the corporations, their shareholders, the officers and the board of directors would be in serious trouble. It is incredibly expensive to tool up to manufacture other products, and there would have to be a market for them. There would be a glut of many of the products, and the competition would be fierce. That is what they have to solve. They have been told by the ones in the higher echelons what has to be done, and they are required to come up with answers. They feel the "ends justify the means," and this is for their survival.

Most are not aware of the true "master" that is controlling the entire process, the one that is pulling the strings. The situation is different with the military. The rank and file are the ones that can make the decisions once there is an agreement about what to do. They are mostly in command of all the decisions that are made. In the case of the US the president, who has a world of advisors, points the way, but the how's and why's are mostly decided by the military.

Even though the meeting that was in progress included a very highly placed group of participants, they were not the ones that make the decisions that control the comings and goings of industry, banking, etc. Due to the fact that the military/industrial complex was planning to move forward, the elite corps decided that it was time to reveal themselves and their plans. So as the meetings were really getting down to the fine points, the real movers and shakers stepped in. Several at the meetings were aware of their control behind the scenes. Some were not aware of them before, but most had been informed that they would be making an appearance.

The Plans of the Elite

The elite family representatives wanted to know what the plans were and any decisions that had been made. Then they said that the group of families they represented had been in control of a large part of the world's assets for generations. When it was explained that the new members were "the money," there were no complaints. The ones that were planning some sort of event would be able to continue their plans, but with a complete assurance of all the things necessary to make it happen.

The representatives listened to what mostly had been agreed on and were somewhat in accord with what had been planned. Then they told the planners from the previous meetings what they had in mind. They disclosed what their plans were, had been for hundreds of years, and would not be deviated from. The ones that had been in the meetings then realized that they had no power at all, that they were being used by the Elite. But they would be well-compensated.

These "power brokers" have had an agenda for hundreds of years. It has taken several hundred years for them to get to this point. (We are not going to get into the off-planet agenda at this time, but believe us when we say that it is real.) These few families intend to take control of everything on the planet; that has been their agenda all along. And they are ready to make that a reality at this time.

Once word got out that something was being planned, the small group at the top then got into the act; so more meetings and discussions began. This group will make all the final decisions. The top echelon has decided to speed up their plans that have been in the planning stages for some time now. Timing was the main question they could not agree on. But some things have happened recently that helped them make a decision on the timing of these events. They are now in the final stages of making their plans happen.

The Basic Elite Agenda (Sananda)

They want more than just the US involved, to accomplish what they have in mind. Their plan is the destruction of property and would involve the killing of a very large number of people. The planners have stated that they will not allow nuclear weapons to be used, but they are not sure that they can keep that from happening. They are aware that there are fanatics that would not care if the world were turned into an inferno of hell in the name of their god.

Then with the chaos that follows they will be able to declare martial law, first in the US and then other countries will follow. Many by now will start to demand that the perpetrators be identified and found. The judgment will then be as expected: Osama and Al Qaeda. So now they will want to go after them in full force. Once martial law is declared and things are under control, the "perpetrators" will miraculously be found. That is when we will step in!

We feel the biggest obstacle, the US, will be disposed of first. That has already been set in motion with the decline of the US dollar. The failure of the dollar will be devastating. The repercussions will circle the globe and have a major impact on the world's economy. The ones pushing for this the hardest are not even aware of the full effect. And that is only the beginning. The power elite will step up and offer to pour trillions into the economy, using whatever currency is necessary to stabilize it, but only if they are in control.

When their offer is accepted and they have control – and this will happen – then they expect to take command of the militaries that are in agreement with their plans. The Elite will be able to do this because they will put their money where it will benefit them, and that is the military. Since many of the high command of the military know little else outside of it, it will be easy to gain their allegiance. They will be told that there is no other way to bring order out of the chaos that is running rampant around the world by then. They expect to do all this without firing a shot.

Taking over countries from within will be the new order of the day. Even though the Elite have been the means of financing wars and furnishing the weapons, they plan to eliminate war because of its effect on the environment. That will be their biggest selling point. Then it will be easy enough to take over the world's armies permanently, again by promising power and money to the right ones. This time they will have to live up to the promises or risk the ones in the top ranks having visions of taking control. Those in power will make it very clear that they have absolute control of all monies to maintain the armies. Then the elite group will set up or put their plans to work for a one-world government. They would assume control and run it like they see fit.

As was said above, when they take control as planned, high on their agenda is to outlaw war. Now, this may seem a little strange because these families have used wars and the war machine to get where they are now. While the chaos is going on, with everyone blaming everyone else for what happened, the families will secretly be putting together an army to take over. That is where the military will come into play. The military leaders that were at the meetings all along will be joined by handpicked heads of military of their own choosing. Those chosen will have been convinced that this is the only way that the complete annihilation of humanity can be avoided.

At this point, the Elite will implement the plan they have had all along. They want complete control of all facets of the lives of all inhabitants on the planet. They believe that they know best and will make sure that everyone knows that. They think that they will have to destroy the control that religions have over the vast majority of the world's population. They want control of all religion.

There will eventually be only one religion – a state-controlled religion. They are aware that a large portion of the population has a need to believe in a higher power. They expect at some point to be that higher power. There are several religious groups that have had similar notions of controlling the world's masses. We of the team have discussed it and feel that they might team up with one of the major religions. How they will accomplish this has not been discussed in detail, but they consider it an important part of their agenda.

What we have given you is from the elite group's own meetings. This is their agenda, what they have been planning for a long time. But as you are well aware, we also have an agenda. Creator has provided some of it already, but we will be filling in the blanks in other sessions.

The Surveillance Team (Sananda and Sanat Kumara)

Sananda:

All right, my friend, much has happened and is continuing to happen. I hope that you are up for the amount of information that we will be conveying to you. It is almost non-stop. You are only vaguely aware of how we get information from the third dimension. We have ones that are constantly sending us information from many different locations scattered around the planet. These beings have incarnated on Earth by a very different means; so they can send information by thought transfer. They are especially trained to do this and have been on the Earth for many years.

Because of their special qualifications, they have been able to work their way into very secret systems of government, industry, and the various religious outlets. We do not consider them spies – the information is only sent to us in much the manner as all thoughts are permanently recorded in the universal archives. It comes as a recording made at secret gatherings such as the ones that we have been monitoring during these past few days. Yes, we have penetrated all the groups on the planet that have been in the process of the preliminary, very subtle takeover of the planet. Up until now, that is.

This leads you to understand that there might be more than one group with those plans. Most plans have been subtle, with the process taking decades, even centuries. In the last several years for this group, the idea of a physically violent takeover has been seen as not practical and mostly impossible. Now it is being accomplished with words in very discrete, seductive ways. But then I said, "Up until now." That will not be the way any longer. These groups will be the victims along with the rest of the population.

In our surveillance we use some electronic listening, as a supplementary system. But there is really no substitute for the "human touch" in the surveillance. We can even get video coverage, which makes it incredibly accurate information. We will need that when it comes to a showdown, which is in the early stages already. That is why we want to get as much of the proceedings in written form by you before the time comes to bring you aboard for your final preparation. These writings will be the last of the things that we needed you to do while still in the third dimension. You will soon see how important that is.

Now to the new transmissions from the ones that are doing the planning. The meetings have pretty well broken up, and now only small group meetings for the final planning and implementation will take place. Secrecy is the main concern now. They have all taken vows such that anyone who even hints to an outside source about what is being planned will be executed in a most violent manner, along with his immediate family members. (Most are family men...and women; there are a few very prominent female members.) Because of that we have to be very careful with our representatives. We don't have every part covered, but our electronic surveillance will take over there.

They want things to move along very quickly at this time. The "restless group" that instigated the early part of the planning months ago is now looked at differently. They are being highly praised because much of the planning had already been done. Because of that they have a good head start. Of course, we are right with them. Much of our planning and preparation has been complete for months also; only a few minor adjustments need to be finalized. In fact, we are a few steps ahead of them at this point.

Sanat Kumara:

We have a being of light that incarnated into one of the very elite families to be a channel of information – our highest source. Unfortunately, he has not responded to our wakeup call and has only just begun to have some doubts about what is being planned. We have covered ourselves, however.

The entire group that is in the process of exercising its plans is very concerned about surrounding this project with secrecy. They are somewhat aware of the possibility that we might try to eavesdrop on their plans. They are certainly aware of us. They have taken extraordinary action to protect themselves and feel that we cannot penetrate their shields. But we have. There is no way that they can keep us from tuning in to their thoughts, as we have done.

This means that we are completely aware of what they are planning to do. The problem is that we will be unable to interfere with those plans. The only thing that we could do was to let them know that we know and see what happens. So we have done that. They are not indicating that they will change any of their plans. They now feel that we will not interfere. It was just a stab in the dark. The only thing they might do is change the timing, but we don't see that happening. As with all matters of that nature, timing is very critical. And since their plans are final, they would be thrown for a considerable loss by changing the timing.

They are now all set to go ahead with their plans and expect them to be activated and completed soon, as we have already informed you. We are not yet aware of the sequence of these events, but I expect that we will be soon, at least partially. They are totally committed to this and have absolutely no qualms about how many will suffer. They have already begun the preliminary work. We also told you that a small, very committed group inside – you would call them a core group – began preparations quite some time ago because they have been committed for the past few years to doing some major acts as these. That made the timing much easier.

Even though the Elite have managed to make it difficult for us to tune into their plans and we may not have all the details, we know enough about them that we can pass it on to you so that you may record it for future reference. They have been trying to come up with a solution that will stop our eavesdropping. They call it meddling and interfering with their free will. All of the dark forces are claiming their right to free will, but it seems to be rather one-sided. They want it for themselves, but they do not want others to have it. They will soon discover that it doesn't work that way.

The Transformation of a Planet

13

More on the Elite Agenda

Sanat Kumara on the Economy

One of the items the elite groups have been discussing is how to avoid what is going to happen to others with the Earth changes. They think that they can prevent being included, that they will be immune to the events. At least they are being told that by their masters. Yes, they have been communicating with others in a manner similar to how you are communicating with us. Except they do not put the words on paper like you do. It is imprinted in their minds to the point that they think the messages coming to them are their own thoughts. As you have been told, some are aware of their masters, but most are not.

The next big issue is about the possibility of another major terrorist attack in the United States, possibly worse than 9/11. There is much talk about it, and it is shaping up as a very distinct possibility. There is information being received from many different sources. No one is saying what kind of attack it will be, just that there is a lot of goings-on underground. We believe that they started making plans as soon as they were so successful in their last attempt.

When I say that there are so many aware of the possibility of an event worse than the 9/11 event, I don't mean the general population, but the military or intelligence agencies and certain ones in the White House crowd. Some were aware of 9/11 before it happened but chose to let it happen and to use it for their own agenda.

You know that you ones in the US are only getting the amount of information that the ones in power want you to have, which is not very much. So multiply that ~~by~~ many times and you can get an idea of what you are not being told and what you are outright being lied to about.

Now I want to tell you what is really going on beneath the visible that seems to be going on. Those things are very different. As has been expected for quite some time, the elite groups plan to bring the economy of the entire planet to a standstill. I am sure that you watch the stock market enough to know that last year the numbers were in the process of going through the roof. As with all of the previous crashes, the plan is always to increase the gains and have as many people as possible put much of their savings and even borrowed money into the market. This has pretty much been accomplished.

The other goal was to get the social security placed in the market (privatized). There has been a lot of talk and a little gain in this but the elite families have decided not to wait for this to happen. You are aware of the death of the eldest of the Rothschilds, and that is going to make it easier for the others to begin the process. Bush has been pressured by both sides – those who do want war and those who do not – about the war in Iraq. The ones in control want it to continue escalating until the collapse of the economy. It appears that this will definitely happen.

As I have mentioned in the past, my main purpose, other than being one of your protectors along with Narno, is to monitor the stock market and the economy of the US as well as the rest of the world. They are all connected because of the multinational conglomerates. Also, I keep tabs on the goings-on of all the governments and rulers of all nations, no matter how small and insignificant. So you see, I am kept quite busy. Of course, as you would suspect, I do not have to be present at these locations and events. I just monitor the events through our technology and sort of put it all together. Then it is viewed by the entire group, so you can know what the adversary is planning.

We feel the falling of the dollar against all other currencies will continue and may even speed up. Most of the economies around the world are in a better position than in the US. But not by much. They are watching closely because they are afraid that if your country goes bankrupt, the rest of the world will follow.

I am continually monitoring the ups and downs of the stock market. A stock market crash is a close second to the failure of the dollar as a cause for a broken economy. The near meltdown that happened caused a panic that circled the globe. The countries that have a daily stock market – Japan, Hong Kong and China – and the European exchange all had major declines, almost record declines. The stock market has been a roller coaster ride the past few months.

The ones of the dark side are reaping large amounts of power from all the turmoil, fear and anger that goes with the panic. They are trying to keep things hidden from the public as much as they can. So far they are keeping the economy from completely falling apart by creating a false atmosphere of "all is good." The fact that they keep lowering the interest rate tells us that it is looking worse all the time.

The economy and the stock market are totally tied together concerning what is going on with the Elite. There are so many severe problems in the US that it is hard to keep up, especially to report what is happening behind the scenes. That is where the important affairs take place, including mistakes, miscalculations, misdeeds and downright stupidity. Right now anything goes. The movers and shakers in the public eye are not the real power, but they don't always do as told. They want to play with the big boys, and they sometimes have the means to make their own deals rather than follow the advice of their superiors.

And the greed... That has been a problem in the past and it is getting bigger. It seems there is always a certain percentage of players at all levels who are constantly trying to figure out how to siphon off some of the monies they handle every day. It is a constant temptation that is hard to resist. And when they give in, they usually believe they can get away with it, and that the rewards are high enough that it is worth the chance of being caught. White collar crime in the US and the rest of the world, in countries large and small, is almost completely out of control.

The amount of money stolen from the large corporations is staggering. But almost the same amount is embezzled from smaller companies and the retail industry. The ones doing this feel anything is fair game. Besides that, the small-time thieves know the odds are in their favor; if they get caught, the worst that will probably happen is they will be fired. So it's worth the gamble. Then there are the ones in many of the governments that are vying for their piece of the pie, and the ones behind those that are really pulling the strings.

My point in explaining this is that the public is the ultimate victim. They are the ones that always eventually pay the price. When you add this to all the other problems with the economy, the effect is very large on the consumer. Most people are saying we are at the beginning of a recession. When fear is this great, anything can happen, and in this case the outcome will not be good. I believe you mentioned before that when a large enough group of people have the fear that is prevalent at this time, they will bring about the recession everyone is talking about.

There is so much negativity being spread by Lucifer that it is a sure bet that something major is about to happen. The little care package that the government is handing out is, like so many experts are saying, too little too late. Just think: that money will have to be borrowed and added to the trillions of debt the US already has. It looks to me as if the "writing is on the wall." Only time will tell.

Several things have already happened that could keep the public from spending as much as they usually do, and that will add to the already stressed economy. The overall rate of inflation, which has been happening over a longer period, is also an important part of the plan. Many are saying the economy is good, in their efforts to sustain the higher spending mode and hide the fact that you are in a slump. But it doesn't look good from our standpoint. We expect the devaluation of the dollar to keep accelerating. That will then be the beginning of the breakdown of the world's economy. There are other factors besides the fall of the dollar that will ultimately bring about a worldwide depression that could dwarf the one that occurred in the earlier part of the twentieth century.

Oil! The ones that are in control in the oil-producing states are members of the elite families that are not in the public eye. They remain virtually anonymous. They were not originally part of the Elite, but because of the importance of oil in the global economy, they were invited into the elite group. This caused much internal controversy, but it was the only way to have a say in oil matters in the US. The elite families also have a controlling interest in many of the global oil corporations, from the oil fields to the refineries. They have investments in all of the key oil fields. Since they possess most of the world's money supply, it stands to reason that oil would play a major part.

This is one of the biggest factors behind the falling dollar. The failure of the dollar is being pushed hard by the elite group we have been talking about. We of the team expect they will be successful. The US, being the world's largest consumer of oil, has to purchase that oil mainly from foreign countries – millions and millions of barrels of oil. The more the price of oil is raised, the more US dollars go to those countries. This is wreaking havoc on the balance of trade and causing the dollar to fall.

So many countries are producing goods more cheaply than the US that it is taking its toll on what this country produces. Few items made in the US are in demand elsewhere. Products the US citizen buys are increasingly coming from foreign countries. The US makes cars and sells them here and abroad, but such a large percentage of citizens buy foreign cars that it is another loss in the balance of trade. And although US and other countries build cars here, the parts are coming from overseas. There is more to the automobile industry than is generally known. Most of the manufacturers are interconnected. It is like one big conglomerate because they own so many shares of each other's stock.

Other factors negatively affecting the economy include the slowing of the housing market; the billions of write-offs taken by the financial institutions for their bad choices; unemployment; and medical costs spiraling upward. We could write pages and pages about healthcare! Then there are the hundreds of billions of dollars being spent on war with no end in sight. Add all of the problems together and you get trouble. When all of this is combined, you can see the enormity of the situation facing the world today. And soon the planet will rebel.

We have found out from our listening that the highly dramatic occurrence that the Elite have instigated – destroying the global economy – will affect a big part of the world's population...at least the most "civilized" portion. The ones that will be affected most are what you have called the middle class. Over the years, mainly in the US and later in Europe the middle class has been the mainstay of the economy. They made the US what it used to be, what it was ordained to be.

The elite ones of the world soon found out that the only way they would be able to achieve a complete takeover of the economy, and therefore the population, would be to eliminate the middle class. This part of the world's population was growing at a tremendous rate and was threatening to upset the Elite's plans. Once the Elite made up their minds to get rid of the middle class, it was only a matter of time.

The Elite want only two classes of citizens: the super elite class – the families that control the bulk of the riches of the planet – and the very poor working class that the Elite will be able to control when the takeover is complete. Within the working class there will be different levels of ones to do their bidding, including the police force that will be required to maintain the Elite's absolute control.

This group will have the best living conditions in the working class, as an incentive to obey their "masters." Eventually, that is what the Elite will require. The ones that oversee the police force will be the most powerful. There will be levels of oversight. The ones of the higher echelons will be fairly wealthy; so they will fight, if necessary, to hold on to their positions. You could consider the group of police overseers and the police force a third class because of their status, but to the Elite they are still in the class that they will expect complete obedience from.

The best way to control the population of this entire planet is first to control the food. In order to control food without causing widespread panic they will continually increase the price of all food products, and that is already underway. Also to impact the supply of food, it will be necessary for the transportation of food to become more difficult, which is already being caused by the continued upward spiral of the price of fuel. This is being done in increments so the population will not notice the effect.

The next step would be the implementation of a government takeover by declaring martial law – taking complete control using armed means. But there will have to be a good reason to convince the military that is necessary. There will need to be a trigger, and that is what is about to happen. The trigger will be the collapse of the stock market. That event will be worldwide, and the panic will be catastrophic. This will result in an immediate increase in the cost of all food items, even water – clean drinking water. And this will then lead to incredibly high fuel prices.

Economic disaster is what the Elite are waiting for to trigger their plans. The military will try to maintain some semblance of order in the population long enough for the Elite to instigate their takeover of the governments. They will do this by promising to stabilize the economy by pouring money into it. But there will come a time when the ones in control will not be able to keep the economy afloat by pouring billions of dollars into it.

Some have tried to prepare for the economic collapse by purchasing gold over the years. And they will make a profit if they decide to sell...if they don't wait too long to sell it. But the consumer goods that you will need in the near future would be the best thing to invest in. Even though gold goes higher as the dollar falls, what can you do with it after the shift when there are no goods to buy?

You were asking if the price of gold is being manipulated. The answer is yes, but then many things are. We see it continuing to rise but slowly, like it has been. We don't see it going to a few thousand dollars per ounce. We will not speculate on whether to buy more gold. At some point it will cease to be valuable. We still say that food and shelter are of highest importance.

The Elite and War (Creator and Others)

Sananda:
Let's talk a little bit about war. Why is the world so obsessed with war? It seems to be a way of life for a large majority of the planet's population and has been for thousands of years. We could give many explanations and go on for hours, but basically it's a "man thing" – the male of the species. In the beginning it was mostly a way that the male could prove himself, to look good to the female of the species. It was a carryover from the animal

instincts that were bred into him through the genes. We can trace that back to the very beginning.

Actually the primates that were the ancestors of modern man did not have that built into them. That occurred when the Reptilians came to Earth and started their genetic experimenting with several of the various primates. The combination of certain ones with the Reptilian race, which was already antagonistic and warlike, was then carried down through the millennia to the present day. As we said, it was mostly a male thing, but sometimes under certain circumstances it can go into the female of the species. Now, as you are totally aware, many of the females of the animal kingdom can be very ferocious, but mostly when it comes to protecting their young.

As the thousands of years went by, there were periods of peaceful times, but some incident would awaken the "male thing" and the wars would begin again. It just got out of hand when some kingdoms or emerging countries or even small communities had more than others. The groups with little would attack and take what they wanted. As weapons became more sophisticated, it escalated. Then profit entered into the picture, and that brings us down to the present.

When a society, such as the one that exists on Earth today, gets to the point where they have to have a "war" to keep the economy going, that's when that society will eventually fail. That began after World War II. Oh, there were supposedly other reasons, but basically after the postwar boom was winding down and the economy began to falter, the idea was conceived to have a little "skirmish." At first it occurred within a nation. But they found a reason to attack another nation, and the war machine was geared up again. Then they conceived the idea of a "cold war": creating an enemy and continually squaring off as if a hot war were imminent at any time.

In the meantime, small "brush fires" were started, mostly in third world countries, until the war machine reached epic proportions. It had to be continually fed or the economy of certain countries would be in doubt. Dealers in the merchandise of death began to flourish. Billions of dollars were made; millions of people were killed. But the countries that were supplying the weapons – made and sold to keep the economy going – also had a faction that was violently against that type of war; so they protested. This sometimes ended in more killings. So on and on it goes.

Some people were angry at the merchants who supplied numerous would-be kings with the weapons they needed to do "ethnic cleansings." How many people stop to think about the fact that there are only a handful of countries that are manufacturing the weapons and selling them to these groups that are doing so much of the killing? Oh, they will say that they are only selling them to countries for legitimate defense, but where do they end up?

Create a problem and then fix it. But the world is changing; things will have to be done in a different manner now. People are beginning to tire of war. They say they have had enough. But some governments and the leaders of the countries are many times ignoring the desires of their citizens. This is happening now. It is not necessary to mention names.

So something new has been added, a new way for the masters of the war machine to stay in business. Now the terrorists have entered the picture. Even though it is still a male thing, this is a totally different scenario. When terrorism becomes as dominant as it is now, people respond by saying, "Protect us from these terrorists." The leaders say, "We will protect you, but we will have to spend billions of dollars for what we need to protect you." These leaders – ones of the military and CEOs of the multinational corporations – have now looked at the handwriting on the wall and have determined that there is a possibility that the voters might be able to shut down the money machine. They are not willing to take that chance.

Gaia:

There have been major meetings of a group of the elite in the military/industrial complex. They have been discussing a plan of attack for some time now. They have been wanting to create another enemy, just in case certain ones in Congress are able to end the hostilities in Iraq and head off an invasion of Iran. They feel that the economy would not survive in a slowdown. They have been geared up to produce weapons of war now for these past number of years and do not want to halt this production. So you see, they have been able to justify more warring and bloodshed to keep the economy moving. They don't see any other solution.

Their plan is to produce another so-called "terrorist attack" and blame it on another innocent country. We can already see the results. Another major attack would be a total disaster for the American people. They would then lose most of what few remaining liberties they have. Speculation is that it would be more devastating than 9/11, and it could backfire and be the beginning of a worsening of the economy. The stock market would more than likely take a real nosedive. And then it's likely that martial law would be evoked by the president. The plan to bring about a major terrorist attack is not speculation. It is for real. They have pretty well done the planning; they have been preparing for some time now just in case the decision was made to go ahead.

Creator:

There was much truth in the information that was given you about the joining up of the one faction of the military with the elite ones, to bring about their agenda by whatever means it takes. As with the missile crises in the Kennedy years, your country was again at the brink of a nuclear disaster. Only certain interference was able to prevent it. But they have by no means given up.

147

The Elite are now beginning to lose their patience with the ones that they feel should have already completed that phase of their plans. They could easily get rid of the ones that they have been depending on, just like they almost got rid of Bush during the 9/11 incident. But they intend to leave everything as is for the time being. They expect that they will have the time needed to complete their agenda and will just wait for the change in personnel after the election. However, they do still fully intend to go ahead with their current plan to totally disrupt the world's economy.

The elite group is manipulating almost anything they can that will cause the dollar to fall. One of the biggest shocks to most of the world is when oil topped $100 per barrel. Though everyone hoped it would never happen, many feared that it would. There is not much chance that the price will drop. After all, that is the biggest tool the Elite have in their manipulation of the dollar. You can now be sure that the Elite are the power behind the price and production of oil. They have more control than most can imagine.

As you are somewhat aware, the dollar is fading fast. It will probably soon fade from the scene as the world's standard of exchange. If they do succeed with that, which they fully expect to do, then there will be a whole new ballgame. Inflation in the US will be rampant and will soon spread to the rest of the world. I do not see the Euro being able to take its place in time. The Elite will then look elsewhere for ones to take the US puppets' place if their military falters.

There has been much written about the 9/11 event, and much of that is now past the point of conspiracy theory. But it has been proven that the ones behind the event are not being exposed by the government. Much is known as to why they planned the event and caused it to happen; so I will not elaborate on that. I will only say that what has been brought forward is close to the truth of what happened. The complete truth, along with the truth of many other events will also soon be revealed.

The Elite and Religion (St. Germain and Sananda)

St. Germain:

I am now mostly keeping abreast of the activities of Lucifer, and that is no small feat! I know that I have not discussed this fact with you up to now, but I will be monitoring his activity much more closely than ever now and will be keeping you informed of his activities. He is working with all the important movers and shakers that are under his control. Some are aware that Lucifer is controlling them because they made deals with him in the very beginning of their careers. He is very persuasive when it comes to recruiting ones to do his bidding.

And then there are the ones that are doing his bidding but are not totally aware of his presence in their lives. He is able to use them in ways that they would never suspect. These ones are very important in his overall plans. There are thousands of them in all walks of life. Some usually think that they are totally in the right and "doing the work of the Lord." Many times they are spreading fear and are very judgmental in their dealings. They can be as dangerous as the ones who are outright in their dealings and who denounce a higher intelligence or Creator. (Now, I know I am on thin ice here, but this is not the time to worry about such things – I am just telling it like it is.)

Some have had their secret rituals and have been hiding in the darkness for millennia, but they are now becoming more open with their dealings and will be even more so in their so-called "pagan" rituals. There are a large number who practice these rituals that are among the ones who control the conglomerates that pretty well rule the business end of all activity on the planet. Also included are high up government officials, those who control the religious institutions, and many of the ones known as philanthropists.

As you can see, they are present one way or another in all the different walks of life, in every phase of civilization. Even though this group is only a small percentage of the population, there are tens of thousands of them, and they are the ones who control most of the comings and goings of the planet.

Sananda:

It is quite interesting that you have been reading and learning much about the various religious beliefs on the planet. As you are finding out, there is much that is hidden from most of the members of many of these religions. The members of the congregation are mostly unconscious and are thought of mostly as non-beings. They are there to blindly give as much as they can be forced to give to support the hidden agendas of the church hierarchy.

There is much that is to be exposed to the "sheep" that will be quite a surprise; overall the results could be very destructive. The members will really feel they have been deceived. There is not much difference between the churches and the business enterprises on the entire globe. It is all about money and power.

Gaia on Weather Manipulation

There has been much taking place that I am directly involved with since we last got together. What I am doing now is letting events have their way while I am mostly just monitoring the certain ones that are orchestrating the weather patterns. In the past I would have made an effort, and been successful, against the ones that are manipulating the weather. But our Creator has said that we are not to interfere with what is going on; so I am just observing and making note of what they are doing.

More on the Elite Agenda

They are still in a learning mode but are close to being able to enhance and guide the weather systems to get the exact results they want. There is so much involved in what they are attempting that they have to experiment to get to the point where they can control every facet of the weather on all parts of the Earth. Even though they are doing this all over the globe, they are not creating anything; they are only enhancing the weather that is occurring naturally. Many storms would not develop into a major event and would otherwise dissipate if they were not being controlled. Their actions will be minor compared to what is coming.

The ones controlling the forces are only interested in power and control. Money isn't the main reason, although that will follow. They already have most of the world's wealth. Controlling the weather is another way they can control the planet. That means they would be able to tell nations – command nations – to do as they are told, or they will cause all kinds of havoc with their new-found powers. It is an awesome thought. And it is nearly a reality.

The Transformation of a Planet

14

An Unprecedented Landing

An Introduction from Creator and Others

Creator:

There will soon be a happening taking place that will answer once and for all whether you ones of this planet are alone in this universe. There is going to be an unprecedented landing of My team – the first landing of this type in thousands of years. There have been several landings over the last fifty to seventy-five years, but they were not allowed to make contact with the Earth's inhabitants. Michael and Sananda (the one known as Jesus) will reveal themselves to the public at this time and explain to the world their purpose for being here.

Soon after the landing I will have My team ready to begin negotiations with the various groups and leaders of the world's governments, churches and certain businesses if they so desire. This will be their choice, but the world needs to know what is coming. I have said this many times, but I want you all to know how imperative this is. My only purpose at this time is to concentrate on you and the other members of the team to complete the final plans for your date with destiny.

I have told you in the past that I wanted your landing to expose the planet to the fact that many other civilizations exist in the universe, and that I wanted the landing to be an event that stood on its own. But Our timing will be coordinated with what the elite group does. As I have stated, we will not interfere with their plans. Their success or failure will be their own doing. They will have backups in case one plan fails. Never feel that they do not have the means to accomplish what they set out to do. They will have access to all the weapons in every country's arsenal, as well as unlimited funds.

I know it is difficult for the team to just sit back doing nothing as this happens, but you are well aware of the fact that we cannot interfere with the Elite's free will. Even I can only do certain things in answer to the call for help. Mainly that is to let people know what is to come so they may prepare.

The elite group's planned incident will accelerate the chaos that is on the verge of happening. That chaos is being suppressed as well as it can be, but this will break loose and open the floodgates that will bring about mass fear. The timing is important, and the landing should be made before the impact of what has happened wears off and total chaos sets in. They do not expect Me to do anything at all about their plans, now or in the future. I have done nothing in the past to prevent what others have done that is much worse (at least to the Elite's way of thinking) than what they are planning. So as you can see, I will not do anything until they act.

This is the reason I have been telling you ones to be prepared to act at a moment's notice – you in particular, Bobby. For as soon as they act, I want you ready to respond immediately.

That means that you do not have much time left in the third dimension. Your usefulness on the physical plane will end at that time, and you have to finish your preparation and be ready to start your part of My plan. Actually you have already started, but now it will be a visible participation. I expect that you will have a few butterflies in your stomach for a time until you complete your full remembrance of who you are. Then you will be in more familiar circumstances and surroundings.

St. Germain:

Most of the things that have been falling into place were in a kind of holding pattern until all the pieces were in place and all the players were at hand – which is now! Everything is in place; all the cycles that our Creator has talked about are in their proper alignment. You, as the host of this galactic event, are just about to make your entrance. As you are aware, you and all of us members of the team are going to make a "grand entrance."

It is certainly going to get the attention of all beings on the planet. There will be no more doubt as to whether the Earth is alone in the universe as far as the existence of conscious beings. There are a few in the scientific community that have the belief that we are not alone, as do many of the world's religions. You are well aware that many of the ones in many of the world's governments are totally aware of so-called "extraterrestrial" beings.

Sananda:

What our Creator has given you pertains to your past, present and future – your entire life up to the present and beyond. My life, however, is known pretty well the world over, although many only have a brief knowledge of who I am. With you it is different. You are only briefly known by a rather large part of the world's population. But even those are not very well aware of who you are. So Creator wants to change that. He wants the vast majority of the peoples on this planet to be aware of who you are even before your presence is known when you make your entrance.

It is a rather interesting situation. You are a longtime resident of this planet, but you will be introduced as the ambassador from the higher realms. That is in fact true. You are more well-known in many other places in this galaxy than you are here on this planet. Actually, as with Me, you are not able to say that you can call any planet your home. We are both truly residents of the universe.

You were only asked by our Creator to establish Earth as your permanent residence for a time. You have earned the right to call Earth home for as long as you like, as long as it takes for us to complete what our Creator has asked of us. But then of course there will always be other desires, other mountains to climb. Even as much as you and I have seen and accomplished, we have the understanding given to us by Creator that we have only begun.

The Landing Agenda (Creator)

First let's discuss our preliminary meetings with the various members of the media. We will invite a pre-selected group from the various media corporations to attend the appearance of our craft. The invited representatives of the different media will be chosen very carefully. They will have already been contacted before the first meeting. That is to ensure that the ones that were chosen would be convinced that this would be an actual landing of an alien craft.

These will be very responsible and trustworthy people. And we will spend time with them in advance to assure them of our intentions so they can assure their audiences of those intentions. We will ask the media to publicize that we are here in everyone's best interests. But there will be those who will try to sensationalize our landing and generate a large following of people who are expecting us to be what we are not.

They will be forewarned that if they release any information of our coming presence, they will not be allowed to come to the first landing of our craft. They will be told that we would also be able to neutralize what they would have released. They cannot tell anyone, including family, friends, and especially their bosses and coworkers. Any leak at all and they will not be allowed to attend the interviews. We will make it known that we will be monitoring them every moment until we make our appearance.

They will be allowed to bring a cameraman, who will also be required to follow all our instructions. After the initial appearance of our craft, they will be allowed to call for the additional equipment required for a worldwide connection. They will be told that they can have them on standby. The media will be informed of the time and place where we will meet with them so they can be there in advance. They will be asked to hold their questions until you ones have finished with what you have to say to the world. Then they will be allowed to address questions and comments to any one of the group. It will have to be brief, but we will set up interviews at a later time.

Now, about the landing... It will take place in the area that was chosen some time ago. I still considerate it to be the most appropriate place. When we arrive at the desired location in our personal craft, but before we make ourselves known, we will break into the local communications systems. We will announce that there is to be a landing of diplomats from various parts of the galaxy. I believe that this will cause less chaos than making ourselves visible all at once.

Then we will be able to cause all military and other aircraft within a large area to be inoperable for a time. So we will suspend all aircraft from being able to operate except for all emergencies, which we will be aware of. Even if there were means to fire on our craft, we can detect and prevent it.

When that is complete I will instruct the commander of your spacecraft to reveal itself to the public before our first encounter with the media. Our personal craft will make a few loops around the area and be visible long enough for all to observe. After we have assured the ones of this planet that these craft are all friendly and will not harm any of the inhabitants, Sananda will display a show of force by revealing a large armada of ships. These craft will be many different sizes and shapes since they will be representing many civilizations.

The world's military will not be happy with our landing. Neither will the larger governments because we did not reveal ourselves to the heads of state first, which is the normal protocol.

They had their chance many times but refused to reveal the existence of other world visitors. Our first meeting will not be more than an hour. During that time their craft will not be permitted to leave the surface, and other craft will be warned not to enter the area. The reason we will make a show of all the alien craft is to show how futile it would be for the military to try to attack any of the craft that are there.

Timing is of the utmost importance. As soon as the word gets out, the military and all the local authorities will swarm in like a horde of locusts. We can prevent their arrival, but we do not want to do that any longer than necessary. That is why our meeting must be completed in no more than an hour. Our armada of spacecraft will be able to easily maintain our protection. We will meet in an isolated area that will not be accessible on foot or by any type of vehicle except the ones we allow. If we do not take these precautions, there could well be an unfortunate incident.

Our biggest concern upon landing, and the one we will address first, is to further attempt to assure the world there is no reason to panic. That will not be easy. We will contact as many of our "light workers" as we can, so they can spread the word that this is a mission of peace and that we are here to warn them of what is coming. Since we will not have time to "beat around the bush", we will get right to the point. We will briefly explain our purpose on Earth at this time and then introduce the members of the team plus the other associates that will attend.

Because of who Sananda is and his past association with Earth, he is the obvious choice to make the introductions. First we will have Sananda introduce himself as both Sananda and Jesus. He will be in the appropriate attire that will be recognized by a large portion of the planet's population. Even those who do not recognize him visually will probably be aware of who he is. Many will rejoice, many will be fearful; many will see it as prophecy come true. Some will expect Jesus to be ready to do battle. There will be much alarm caused within the ones that expect Jesus to take his place as the "new king." But Sananda will quickly address that expectation as he explains why he is here: what he is here to do and not do.

Sananda will then introduce the other dignitaries and tell where they are from. The team will be introduced last, with you being at the end. You and Sananda will then relate to the world why we are here and who you represent by informing everyone that you are My representatives, here to relay My message to the world. You will reveal the reasons you are here to the entire world at the same time. Then you will briefly discuss your plans for the coming days.

You will assure the ones of this planet that you are here at My request because I have been asked by many souls to intervene in the situation occurring on this planet and that I have been waiting a long time for the ones of this planet to finally feel they have had enough of what is going on here. They have finally asked for My help with enough sincerity for Me to respond. That is the only way that I can intervene. Now that I have, most will not like what I have to say or what is about to happen next. This is the time that I want you to also tell them that I am going to make an unprecedented appearance on this planet.

When you have finished there will be a somewhat orderly question and answer session. After the questions we will excuse ourselves and end the meeting, but not before letting the media know that you and Sananda and other members of the team will be open to private interviews.

At the close of the meeting you will then return to your main craft and prepare for meeting the various ones that I will designate.

I am sure you are aware of the responses we will receive from the world's religions. That will be most interesting, to say the least! I suspect that many will demand an audience with you and Sananda. Each one will probably expect to be respected as the one true religion; so we will have to treat them with kid gloves, but we will have to be firm above all. That is the term that I will use in dealing with any and all governments, religions, businesses and individuals.

I wish that I could say that this landing could occur without causing panic and perhaps widespread fear. But I do not see that being the case. Sananda and I will spend as much time as possible trying to calm the population.

A Message to the World (Creator)

This landing event will stand alone...to make a statement. (Actually, there will be multiple statements.) Upon our landing and brief show of force, the entire planet will learn the truth about one of the biggest lies and cover-ups: that Earth and its inhabitants are not alone in the universe.

And once that becomes evident, it will be very difficult for the ones in control to keep other truths hidden. The public will begin to demand the truth. World wide, intelligence groups will stand accused. Those who have been trying to expose them will seize the opportunity. I see no way to avoid a serious situation. Much anger and frustration will be vented with words, and some extreme violence will also be unleashed.

The next statement that will be made is that all of Earth's population is ruled by lies. The higher echelons of government are only the starting place. All governments are based on convenient lies. The sad part is that many people know that is so, but they accept it or ignore it even as the lies get bigger. Nations built on deceit, especially those that are supposed to be "of the people, by the people and for the people," are doomed from the start. Although some are worse than others, I see not even one government that is not based on deception.

It seems to Me that your entire civilization is deceptive. Untruths or partial truths are told by most of the population and believed or at least accepted by the others. It begins with the youngest with "little white lies." You have to lie to avoid hurting someone's feelings. A lie is a lie, no matter how you look at it. Is a lie ever justified? I will let you ones decide. There are lies in the business world. One falsehood leads to another until it gets out of hand. Does it ever stop? How about the lies in your various religions? You would think that would be the one place you could count on for the truth. But that is not the case. That could be a rather delicate subject; so I will not dwell on that now. But that doesn't mean that I won't sometime in the future.

One of the most important statements that I will make through My representatives is that the entire planet must at last know the truth of who they are and where they came from. That is why My team is making themselves known at this time. Now this will step on a lot of toes and hurt the feelings of many. That cannot be helped.

The final statement is not the least important just because it is the final statement. It may well be the most important. It will have the most impact because it concerns the whole population, as well as the planet itself. It is the final warning. You have known and in many cases have been warned that the future of this planet is in doubt. The humans of this Earth were put on this planet to oversee all the other beings, both animate and inanimate. But instead of taking care of them you have done the exact opposite. You have abused them to the extent that many species are now extinct. In some cases extinction is the natural evolution of things, but not to the degree that it has taken place on Earth.

The amount of pollution that is now taking place is in danger of overwhelming the planet so that it is in danger of dying. And that I cannot allow. There are two reasons why I can overrule your free will status, and that is one of them. You have been given many small warnings; you are getting them every day and that will continue. But now - know this! This is the final warning (a book with that title will soon be available to all): the Earth has now reached the point of no return.

She herself is about to carry out the final statement. It will not be pleasant; it will be very drastic. Now, this is not My judgment against the inhabitants of the planet. It was brought on by all of you. She is going to shake and roll as never before. I have given My consent because what she is about to do is in her own self defense. I will not interfere; it has gone too far. It has to happen and very soon.

Very soon each person will need to look deep inside and find out what they must do to cleanse themselves. At the request of the one writing this message (the author of the above-mentioned book), I have recently sent out a wave of awareness to every soul on this planet. The message was given subconsciously to give you one more chance and to tell you what you must do. It was mostly ignored, and it is now out of My hands.

We want our appearance to be as dramatic as possible. Yes, "dramatic". Are you surprised? There will be much panic. After all, look at what is occurring or about to occur. How can we possibly prepare the inhabitants of this planet for what is to come? I know that you have been concerned about that for some time now. It will not be easy; that is a certainty. Because things have come so far, what we need to do is give the people hope - hope for the future, even if the immediate future is fairly grim. At the same time, they will be made aware of the information in your book and that we are here to help them through the devastating times ahead. As you have stated in your first book, the inhabitants have been warned many times, but they have ignored the warnings.

But what about the future down the road or beyond the horizon? There is an absolutely glorious future there. So, Bobby, we have to give them hope, and that is what I intend for you ones of the team to give to the entire population of this planet. That hope will come from giving them the truth about who they are, where they come from and where they are going.

They will also be told the truth about the lie they have been living. And since they have done very little to find out the truth, the ones telling the lies have just continued, because it was so easy. But now it is time for the truth to be known. The ones that have been telling the lies will have to make the truth available to all. They have been overstepping the people's free will, even though that free will was turned over to them. And for that, those that have taken advantage will have what you could call instant karmic payback.

Meetings with the World's Governments (Creator)

We will first arrange to meet with all of the countries that have an interest in knowing our reasons for visiting the Earth at this time. It will be their choice as always. So many of them have had their chance and would not take it.

We have talked about the possibility of meeting in the United Nations building, and that is still My first choice. I have looked at it, and there will be no difficulties with that choice. As I have said, I will look at the inner being of the ones of each nation to designate which ones that I determine would be amenable to talking with you. I have seen that most will want to meet. Explaining our purpose for being here will be the first item on the agenda.

Again, all the members of the team, plus all the supporting ones, will be at this meeting and will be introduced again to the world. Greeting the Earth's inhabitants will be full of surprises. After the introductions you and Sananda will make your presentations to the interested leaders of nations, the world's religious leaders, and the corporate and military/industrial conglomerates. We will explain that this will not be a question and answer session; it could get out of hand.

When the main session is finished, we will again meet with the world's media representatives. This will be an extensive undertaking. Not all of our team will be present at this occasion. I will reveal to you at a later time who will be present. I have also not discussed with you what you will present to various nations and all others in the way of explaining why we are here. I will not do that until after you come aboard and we can get a reaction from the entire team.

As I have told you, I will be at your side guiding you as you meet the world leaders. And Narno and Sanat will be there as your protectors. We have not mentioned this before, but you will have considerable powers that you could respond with, but that will be the job of your protectors. It will be best if you do not respond except in an emergency. The big question is how to present to the world what is about to happen to the planet almost immediately after our presentation.

We will have previously told the nations' leaders that after the main meeting and after we have returned to our spacecraft, we will then contact each nation again and arrange meetings with those that desire them. But we will not meet with each nation separately. There will not be enough time, and it would not gain anything for them or us.

By that time we will have arranged a schedule of meetings for sets of neighboring countries. Each will be allowed equal time, and there will be no arguments or quarreling. We will firmly state that this will be a peaceful event. Each country will be allowed equal time and will be treated with equal respect. The larger countries will not get special treatment. All will be treated as equals because that is the way the Creator sees all of His created beings.

Since timing is critical, who to meet with or not will be a sensitive matter. I see you thought of employing your other selves. That would not be appropriate for this delicate situation, but everything will work out as expected. After all, you will be the one to control the agenda. All will have to understand that you are here at My request, doing My bidding.

Responses to the Landing (Creator)

You ones of the team have been concerned about what the Elite will do and when as to their plans. After we make our appearance during that event and have made the world aware, the elite group will at that time feel that they will have to make their move. They have wanted to for quite some time but could not conceive of a correct time. This will then force them to reveal their intentions to the world. This will be the beginning of the showdown between the forces of light and the dark side. You are wondering why they would pick that time. Just remember, they are being guided by the master himself, Lucifer. He is ready for a showdown.

He will now put all his cards on the table and will call forth all the ones that he has made promises to – the ones that have basically given their souls in return for riches and power. They will follow his commands or else! They will have no other choice, except to come to Me. Now, this will not happen overnight, but it will very quickly escalate. It will begin to take place as we reveal our purpose for being here. Also, as I have said many times, I would prefer that it could be different, but the ones of this planet have brought this on themselves, and this is the way it has to be.

Now, I see that you are wondering why the dark forces will continue with their plans after you make your landing. It will take place while you are having your talks with others, which means that they will not heed My warnings. They will do this to make a point: that we will not interfere with their plans. And of course we will not. After much truth is revealed, the population of this planet will believe that they have been manipulated long enough. The Elite's agendas of the past will begin to be exposed to the light of day, and the people of most of the countries will make quick work of the plans of the Elite. All of the plans of this group will then be of no use to them.

In the meantime, you will begin the next phase, which is to warn the ones of this planet about what is to happen next. When you start to give the warnings of things to come that I have given you, the beginning stages of Earth's cleansing will begin. There will be many events taking place that will climax with the shift. That is a given.

Even though there will be much turmoil, some will try to go ahead with their lives. Many will listen to your warnings, but you will have to be constantly reminding them, as will Joshua, that time is running out. So many things will be happening that most will then believe what you are saying. That is when the chaos will really start. That is also when I will make My presence known and will then answer their prayers. They will have already been made aware of what that will require.

The authorities will make an extra effort to keep the Internet alive. I will intercede with that so that everyone will be kept informed about what to expect. That means that your books will be their best source of information. Most religions will not be looked to with much faith. At some point I will then bring forth My other means of preparing all for what is to come.

So it will be a "double whammy", or even a "triple whammy." First your warnings and books, then My making My presence known, then My "secret means" for all the ones that have not been warned by other means. I will reveal that to you ones at the appropriate time. You ones, My Special Forces, now have close to the agenda that I have planned. I say "close" because there will always be a few minor changes. But as all of you are well aware, it is too late for any major changes.

As you watch the daily events around the world, you can get a very strong feeling that the end must be near. People are correct in their thinking, but that doesn't make the possibility less frightening to the world's population. Nothing will relieve the panic that is building. There are more outbreaks of violence; more individuals going "over the edge" and killing family members or groups of strangers; and more revolts due to political maneuverings and phony elections than ever before. People are getting angrier and more frustrated world wide. Let us hope that a reasonable amount of the world's inhabitants remain stable enough to help others in need in times of crisis.

As a rule, the majority do respond to the needs of others. Many respond with money, but soon that will not help. Food, water and shelter will be the main concern. Many will believe it is premature to talk about these matters, but I say now is the time to begin to take action. The government spends too much time studying the problem, forming committees, and setting up action groups before they can begin to take action, which should be taken immediately. That was proven with the aftermath of hurricane Katrina. In the beginning, individuals and local groups respond. The government's biggest response is to send in the military to maintain control. That is necessary, but they tend to overreact and can sometimes cause more fear and disruption. There will have to be more cooperation if the world is to survive what is coming.

I have been aware that you have thought that I would not be very involved in these events. You figured I would sort of be on standby. But you are beginning to see that I will be very much involved in all of the happenings. But, as I have told you, I will not be able to make anything happen without your personal appearance on the planet – a face to face situation. I will have to do many things through you to make this event real to the population of the world. They will not want to believe what is about to occur, not only with regard to the shift, but what each being needs to do personally. They have been led to think "seeing is believing"; so I will have many tricks up My sleeve to enable them to see. They will not be able to not see. Part of My plan, as I have said, is making Myself known.

15

Michael Leads the Negotiations

An Overview from Sananda

You are somewhat aware, as Sanat has mentioned, of some of the things that are taking place around the globe. We are monitoring most of the events that are taking place. It will be an exciting time when you can see the gadgets that we have to do that monitoring. It even amazes me when I watch and listen to so much going on. I know you are having a problem believing that we of the higher dimensions use such equipment. But believe me, we do. We have been doing this for years and instantly replay any event that has happened at any time. This is not only recorded audio, but visual of any event of any major happening. So there can be no denial by anyone of what has taken place.

This does not mean that we record every little thing, but we are recording the events and deeds that have a bearing on what has taken place and is taking place – what effects the environment of the planet or the destruction or manipulation of the inhabitants. Let me explain a little. When a being said to be of the dark side has a thought, that thought is instantly sent out and is then forever recorded, just as for all beings. Since there is a different energy with those kinds of thoughts, they can be singled out. Now you must know there is no judgment with those thoughts; they are just recorded. And the individual that sent out those thoughts or deeds can then be presented with them if necessary.

We will now discuss further what lies ahead of you. From the moment that we make our first appearance on the surface of the planet, things will really begin to happen. After we land and complete our first talk with the members of the press to explain our presence here (which should take no more than an hour), we will return to our spacecraft for the final preparation of plans to meet with the various heads of state and others.

Our first meetings with any Earth groups after our landing will hopefully be at the United Nations headquarters in New York, as we have already stated. But, Bobby, these have not been set up yet, at least not at the UN. It was not a demand, but Creator strongly suggested that is where we should lay our cards on the table to make our purpose known to the entire population of the planet. We will explain that we have no choice other than to come right out and tell the truth about Earth changes they can expect to happen.

The response we get from the first meeting will determine how the plans are finalized and carried out. Then we will begin contacting those we intend to talk to by our own electronic methods. We have the means at our disposal to interrupt radio and television through the Earth's satellites. And we can directly communicate with any individual or government through their own telephones. We will then arrange meetings with those who have been designated by the heads of state and national governments, business executives and leaders of the main religious orders.

Even though I said before that I would only make one appearance, Creator has requested that I make at least two more. I briefly mentioned that you and I and Elijah will make an appearance together in the beginning. Elijah's presence with us and inclusion in our plans will result in a significant event.

The changes will begin to happen soon after we finish our meetings with all the ones we want to meet with. We feel that the three of us will be able to support each other at the first meetings, where we will make an appeal to the nations of the world to cease fighting among themselves. They must be able to understand that they will need to be ready to support each other in the days ahead. They will need all the comfort they can get.

It will be of great concern to many – devastating for some – that I have not come to fight their battles for them or take them up to heaven and save their souls for eternity. There will be an even larger group that will say, "I told you so." It will be that we will face problems coming to us from every direction. So, Bobby, are you beginning to get a feel for what is facing us? I see that from your standpoint it looks almost hopeless. We feel that it is not hopeless, but it will be a staggering situation.

As you know, things are really about ready to explode, and we just want to make sure that you are ready to accept what is coming and what our Creator has asked of you. I know, I know. You have been asked this over and over, but it doesn't hurt to ask you again if you are okay with what we know is coming and you suspect is coming. There is a difference between saying that you will be okay with what is about to take place and experiencing the actual events themselves. It is not going to be a pleasant experience for anyone, and we are mostly preparing you for the worst.

I know that most of the ones on the planet think that is a really bad way to look at things. That thinking was okay several years ago. But since the same ones, along with the rest of the population of the planet Earth, have not taken any steps to change things, at least not enough to amount to anything, about the only thing left is to be aware of the "bad" things along with their positive thinking. It certainly does not do any harm to continue their positive thinking, and for some it will definitely help.

I'll say again that you feel like you are the only one that feels for sure that the stuff is about to hit the fan. That's not easy for you, and that is one of the reasons that we keep asking the same question over and over.

Also, of course, we don't want you to feel abandoned. You have the support of the entire universe; so don't feel abandoned. We all love and support you to the maximum.

Setting the Ground Rules (Creator)

After the initial contact with the ones that have been chosen to meet with our delegation, we will return to our spacecraft for a short time. We will then begin to contact the various governments and institutions we wish to have meetings with. Some will just write us off; some will only send emissaries. But most will be anxious to begin dialog with the first beings to make themselves known to more than just a few representatives from selected governments.

My main reason for not following protocol is known by some, but they don't know all the reasons. Many of the world's governments have been contacted by off-planet beings, but they were not willing to let the population of Earth know of their existence. But mostly it is because many of the ones that are currently heads of state are not the ones picked by the people. Even if they have been legitimately elected, too many of them have their own agenda. They do not have the desires and needs of their countries at heart. Many of the emissaries will only be at the meetings out of fear. Their fear will be caused by the agendas in their own countries.

Because I have asked others to join you at times, I will now discuss some of the arrangements that I have vaguely mentioned before. None of these ones will accompany you at all times, only when you attend large congregations, like when you and several of the members might come before the entire assembly at the United Nations. I perceive that that will be your first contact. There will be much anger displayed at that meeting because I do not intend to allow the so-called powerful nations to dominate this meeting. All countries will be treated the same. Even domineering dictators will be treated the same as all heads of state. The actions of all these ones will determine how we approach them as we begin to meet with them individually.

We will make it clear that there is a timetable that has to be met because of what is to come. It will also be made clear in the beginning that there will be no compromises. And because of time limitations, the discussion will not be allowed to drag on and on as usually happens between states. Time constraints also dictate that we will meet with regions of neighboring countries. Although many close neighbors are hostile to each other, we will deal with that at the time of the meetings. After we have met with the government, religious and corporate leaders more or less individually, we will hopefully meet at the UN headquarters again.

We shall now begin a new important discussion on more of the facts that you will need to know when you begin your negotiations with the various groups and leaders of the world's governments, churches, and businesses. We have talked about this briefly at the beginning of My dialogs, before you were aware that the content might not all be correct. Now I will be able to be more precise.

Probably the first ones arranged to meet with you will be lower level officials. You will have to quickly determine if they are just that – ones that will have no authority to make meaningful commitments. If and when you establish that is the case, with My help, we will bring the meeting to an end. We will have no time to spend with those ones. We will tolerate it for a short time and then we will make our point: we will deal only with the ones capable of instant and binding decisions on behalf of their organizations.

You will be talking to some of the most intelligent beings on your planet. Once your intentions are known, as well as the purpose of the meetings you will request, you can be assured that they will watch your every move (your body language), and they will try to anticipate your very thoughts. In most cases we will be able to block their focusing powers. Minor governments and business conglomerates may not have access to ones that are able to discern your intentions. If they are the higher level officials, they will have access to ones with strong psychic abilities that know the black arts quite well.

If by chance some of those psychic beings are what you call psychic vampires, they will try to suck your energy out of you. But you will have the most powerful of all protection: My protection. Through Lucifer, they will use every trick in their bag to try to penetrate your shield and defeat you. Of course, Narno will be at your side, and you will have the help of Sanat if it is needed.

In many cases you will be alone, at least from the physical standpoint. But you will always be aware of My presence and My thoughts. One of the reasons you will need to be attuned to My thoughts is because you will mostly be relaying My thoughts and instructions to all concerned. Do not feel that you will only be a messenger boy; that is far from the truth. I can only give you messages to relay.

You are the one they will be watching and listening to. It will be your expressions, your emotions, your ability to be very stern in your expression and your "words of wisdom." I have complete trust in you and how you present yourself. What you say will have to be very much to the point, and there will be no time for you to be Mr. Nice Guy. That means that your facial expressions, your words, and your thoughts will have to be in accord.

So you see, the negotiations are very involved. There will be those who will vie for the chance to present their case, saying why they should be the exception to the rule (that everyone must take their turn; none are more important than the others), even after they have been told there can be no exceptions. Despite what I am telling you now and what will be made known to you at that time, there will still be unknowns. So I repeat: you are not just an errand boy; you will be required to make snap decisions at almost any time.

Even though you will be given much information about what to expect concerning these meetings, just know that the unexpected will also happen. I have been talking as though you would be alone with these ones. That is true for some of the time, but part of the time others on our side will be present. I am speaking of physical beings here, but you will have many unseen helpers at all times. You will never be alone for a moment. Some will be sending their thoughts to you from our craft and your guides and protectors, Narno and Sanat, will be at your side – Narno at all times and Sanat when needed. When necessary, Sanat would be there in an instant.

You understand now why you have been asked many times if you were comfortable with what you have to do, do you not?

On another topic, what about the rest of the team? What will they be doing if they are not going to the surface? Believe me, they have been busy enough keeping up with what is happening on the surface of the planet, and they will continue with that task. It will be important to know what is going on at all times so that you can tell the ones you will be conversing with that you know what they are planning. That will make a big difference in your negotiations. So the other team members will be giving you information as it is needed.

This whole plan is not going to be what the ones on the surface have been expecting. But it will be a modern situation that the ones on Earth can relate to. It will be like negotiating a business deal, or negotiating peace...with a twist. Don't let this overwhelm you. When the time is right, you will be completely prepared with all I am placing in your consciousness. And I have complete faith in your ability to do what I have requested of you.

Free Will is Imperative (Creator)

It is good that you are now able to receive each message without feeling that you have to judge it as being truth or disinformation. However, you will still need to use discernment, whether the information comes from earthbound beings or from the higher levels of existence. After all, you must be able to determine the truth from the ones you will be dealing with.

When you meet with the various groups and individuals on the surface of the planet, there will be occasions when you will be on your own with these ones, and those will be critical times for you. It will be important to be aware of what they are relating to you. I will be there with you at all times, but they will not know that. I will be guiding you as to what My commands are. And they will be strong commands because that is mainly what they will respond to.

However, they cannot be compelled, even by Me, to follow My commands because of free will. They will not be punished; I will not strike them down on the spot. But they need to understand that it will be in the best interest of the ones on the planet if they follow My commands. It will also be in their best interest.

It is imperative that you remember free will as well. Trampling on it cannot be allowed under any circumstances. The result of that would be catastrophic to My plan. I have not emphasized this to you as firmly as I am now. Even though many on the planet are guilty of the mismanagement of the environment and have the planet on the brink of destruction, I still cannot infringe on their free will.

I stress this because I have been asked to interfere. I am going over it again because I want it to be as ingrained in your consciousness as it is in Mine. Although I am bringing this up with those who will be with you and guiding you, I am making the point more strongly with you because you will be the one responsible to see that nothing infringes on free will.

There will in fact be multiple reactions from beings on this planet, and you will have to decide how to respond. This is the area where the two newest members of the team, along with Me, will be of tremendous help because of their special abilities. It is not that I do not trust you to make the correct decision. It is that it will be easier on you if you have help. It will be important that you make these decisions without any pause in your conversations with all the ones you come in contact with. You are starting to see the importance of having a team, are you not?

There is the remote possibility that some of these individuals may do something you are not expecting; they might try to surrender their free will to you. That might happen because they could be overwhelmed by being in your presence. After all, they have never been in the presence of an archangel before, and they have not had suggestions relayed from someone who they are told is their God. They will be even more awestruck when other team members are there with you in the physical.

You must not allow anyone to give up their free will to you. That is what many are doing with Lucifer, whether they are conscious of it or not. We both know the problems that it is causing because of the disastrous results that follow. And of course those results are in Lucifer's favor in his bid for total control.

You might think, "What if they surrendered their free will to me and it was put to good use, like the cleansing of the Earth?" That is not what I intended when I gave all free will. Regardless of the use of one's free will – to good end or what one might believe is "evil" – it might not be what the person who surrendered their free will intended. And what one person thinks of as good might be another person's idea of evil.

So the one that accepted someone else's free will would have to be judged as to its use. Who do you think would do the judging? Do you think I might judge? I think not! The person who accepted another's free will would eventually have to judge themselves and deal with it. This will not sit well with some who feel it is all right to accept what is offered. Sooner or later the price must be paid. And the way time is accelerating, it will be sooner rather than later. Bobby, I can see that you would not want to take someone's free will even when offered, but I want everyone to know that in the past, what was rationalized as being okay is not okay.

Dealing with what I have asked of you, when the time comes, is all you need to be concerned with. When the rest of the team is giving you information concerning a group or even an individual, they will be doing so with My approval. They will not tell you anything, nor will I, that is inappropriate. In other words, they will know whether it is an infringement or not.

Know Your Adversary (Creator)

You know about Lucifer. Soon you will be exposed to the other major players – those that have accepted Lucifer's "gifts" and those he has made promises to. Many have accepted. I am speaking about the ones in the higher realms of his soldiers – his top aides that are totally committed to him. They believe in him and the promises he has made, especially those closest to him.

Sananda's information was complete about what the ones in control of your planet have in mind for all its inhabitants. But that group is only partly aware of what will shortly happen to the planet. They have closed themselves off from any messages that I have sent to all beings on the surface regarding what is necessary to survive the changes. They have only one desire: to serve their master, whom they are totally aware of. Lucifer has handled his minions well. He has managed to block any remembrance of their true selves. He has told them it is their right, their destiny to take over this planet.

And his message is being broadcast to everyone on Earth; so each being is hearing that he has a right to control those around him or her. If anyone has even the slightest desire to dominate others, Lucifer is able to begin taking over that individual's consciousness. Once he gains a foothold, it's easy to maintain his power over that person. And the longer he has that power, the harder it is for the individual to break free from it.

As the person's need for control continues, it becomes all-consuming, and the universal consciousness has no choice but to manifest that desire. It may manifest in a number of different ways. The individual still has free will, but Lucifer only has to plant the thought. If the being is receptive, half the battle is won for Lucifer. That is what is happening to people at this time. From My point of view, the planet is continually growing darker.

The ones controlled by Lucifer will not be happy with only a single being giving them My ultimatum. That is what it is – My ultimatum, the final warning. Many will bring up the fact of free will, saying that I cannot interfere with Earth matters. But we will counter that I have been asked by many beings, through their prayers, to intervene in the affairs of Earth. I have heard that call, and I have responded. We will make it very clear that what is to happen has been brought on by the people themselves. We are only here in response to their pleas, and to explain that we are also here to give a warning and then help the ones who have asked for help.

Soon it will not be so easy for ones in power to continue their lies. The longer they persist with their lies, the harder it will be for them to dig themselves out of the hole that gets deeper with every lie. We will strongly point out that fact. They will be told emphatically that it is time for the truth to be told to all the inhabitants of the planet.

In the various government and religious institutions there are many of the light forces from the higher planes who have been placed in high positions. In some cases you have placed some of your other selves there, and sometimes at My command, to bring forth the truth – our ace in the hole. They have been waiting for this time and will awaken to the calls of our forces. They will come forth in small numbers at first; then many will follow. There will be a domino effect as one after another comes forth. Most have been waiting for the right time to reveal the truth. Some have lived in fear and could not come forward. But now all may say what is true without fear. It will soon be impossible for the lies to be believed any longer.

These will be tough opponents, coming at you from all directions. To them, this is not a game. They are in it until the bloody end. They feel they have nothing to lose and everything to gain.

But remember, you are not playing by their rules; you will make the rules, the rules I have given you. Your opponents will not be very receptive at first, but they will quickly be made to understand they have no choice and that you will be relaying only My thoughts to them.

We will be telling them that what is to come is due to the actions of the entire population of the planet. We will not demand anything, but will not argue with them. We will simply tell it like it is. As a longstanding citizen of the Earth, you have the right and a duty to do this. You know their plans for the future of Earth and that those plans would infringe on the rights of others.

You are judging their actions, not their being. The use of free will does not include suppressing the free will of another unless that being has granted permission. Your opponents will then try to say that freedom was given up in exchange for favors or the promise of favors. But they are guilty of infringement if they have controlled just one unwilling person on the planet. And you can assure them that many of the oppressed have called on Me to come to their rescue.

I have sent you in response to their cries for help. You are My personal representative to whom I have given the power to do what is necessary to free them. Millions have been praying for a very long time, sincerely from their hearts, including you. Make that clear to your opponents. Tell them they are looking at one who has asked for My help and who represents all the others who have asked for it.

Just know this: the more you know about your adversary, actually adversaries, the better. There are many that you will consider adversaries, even though Lucifer is number one. In the final moments of the game it will be just you and Lucifer...and Sananda. He has played a big part in this, but largely from another level. The three of you have been the main players in this game that has gone on long enough.

As the "referee" I am now calling for an end to it. It has been the only game in town, as you ones say, and I will soon be explaining the rules for the next game that will be starting shortly.

But first we will have to bring this one to an end. And as I hinted at in an earlier writing, there will be a surprise ending.

On Free Will in General (Creator)

Now back to free will, which is closely tied to judgment of others. What the others are bringing forward to you in your writings is to get you to be able to discern the fine line between what is allowed and what would be infringing on others. Mostly we are talking about the fact that all beings have the right to do and think as they desire. However, this does not apply when what they do is harmful to another being. This not only applies to humans and any other conscious being in all of the universe, but to the animal kingdom and all the various types of lower beings as well. It is a little different with the plant kingdom. Nevertheless, they have an awareness also; so I would suggest you not wish them any harm either.

If you are observing that someone is doing harm to another, whether it be physical or mental or degrading them in any manner, you are not interfering with their free will if you suggest to them that what they are doing is not in either one of their best interests. I am not saying that you must bring your observation forward so they may be aware of what they are doing. It will be your decision as to what you will do. But if what is being done places that one's life in danger and you do nothing, you may have to resolve that later. I am also not suggesting that you put yourself in danger. As with everything else, it is your decision. There will be no karma involved for you.

There is a very fine line between interfering with one's free will and the fact that from the perspective of higher dimensions there can be nothing kept from other beings. All are aware of others' thought patterns. Just knowing something about what others are doing is not infringing on their free will. But that does not mean you will shove it in their face and accuse them of wrongdoing. It is only in the lower dimensions, mainly in the third but in some cases also the fourth dimension, where the problem of interference occurs. When beings are able to move into the fifth dimension, the problem ceases.

As I have mentioned in another writing, it will take some time for ones to adjust to being aware of others' thoughts. It is not a situation where all thoughts from all beings are coming at you from all directions, because that would be utter chaos. That is mainly the situation that you will have to become adjusted to. It is only when you are observing someone that the inner connection takes place.

If you are observing more than one being's thoughts, it can be a little difficult. You will get to where you can handle that kind of situation up to a point. The longer your consciousness is in the higher planes, the more adept you can become. Just be aware that you never cease to learn. That is a constant. No matter how long you are in those planes or how high you go, it is a continuous learning situation; however, it will be somewhat different from what you will imagine.

Now to the real nitty gritty, as you say, of what is to take place. How far can we go without that infringement? I have been asked by many for help in what is now taking place and has been for many centuries. *What has been going on is infringement on others' rights, and it has gone on long enough.* **I will make it very strongly known to the ones of this planet that will soon cease.**

Because of the Elite's infringement on others, I am now free to make this a very strong statement. Even though we are here to observe, we (when I say we, that is Myself and all of My team) are also here to let the downtrodden of this planet know that I have answered their call. I will also very strongly emphasize that you are My personal representative, along with the others of the team that I have assembled, and that you have full authority to act on My behalf at this time.

It will not be My judgment as to what has already taken place, nor My judgment as to what is about to take place. Even though I am responding to the call of many on this planet, all on this planet are responsible and will have to be *fully accountable* for what has been happening on and to this planet. She will not accept this any longer and is about to respond.

So I will say again, this is not a judgment from Me. What is to come is in response to the actions of all that have had an existence on this planet for the last several thousand years. I have nothing but the unconditional love you have only heard about; for none on this planet, with few exceptions, are aware of the true meaning of unconditional love.

I also have full respect for all of My created beings. You have just been led astray by others, not the least of which is Lucifer, for he is very busy on this planet at

this time. I am extending My hand to all at this time for the help that they will need to survive the coming catastrophe. All you have to do is ask from the heart.

Your Loving Mother/Father Creator of All That Is and all that will be

A Postscript from Creator

When I told you that I was moving into a higher dimension, just know that there was a very specific reason for that. From where I am now it is impossible for any being, from the dark side and even from the light, to listen in on what I am now communicating to you ones of the team unless I specifically want that to happen. And since I am now communicating only with My chosen team, that means that no other energies can break into our means of communication. That is very specific. You would not be able to receive My communications if your energy level were not such that you could receive them. There are others at that level on the surface of the planet; however, I am not having any dialog with them at this time.

I will be guiding you in all your meetings with the various groups and individuals. After that I will cease to be involved with the events that are to follow, but I will still be available to you and all members of the team at all times. I will then begin to be available to all as they come to Me for the help they will be seeking. Even before the shift occurs you will be coming from and going to the spacecraft quite frequently for conferences with Sananda and the rest of the team.

This will be the time that the team will really begin to be much more involved. I now feel that Sananda will play a more active part at this time, but it will be his choice. Nonáme, Sanami, Joshua, Elijah, Sanat and Germain will all have important roles. The remainder of the team will continue working behind the scenes to keep you active ones completely aware of what is going on.

You will be in communication, telepathically and sometimes in person, with the ones on the planet as they awaken to their part in all this. Through you, I will also be available to the ones who have not felt they could come directly to Me. You will be very active at this time, as all will know that you are the one for all to look to for guidance.

The Transformation of a Planet

16

Nonáme's Introduction

First Session

Hello, Bobby. I have not communicated with anyone on your planet before you. I wanted to be what you call an expert in your language before I attempted. So now I feel comfortable with your language (English), but I do understand that yours is more what you call Americanized. Is that not so? Good. I have listened in on your conversations with the others – eavesdropping I believe is one of your terms. Yes? Good. I especially enjoy listening to you and the One you call Creator. We do not refer to Him, also another term you use, with a personal name. To us He just Is – the all that exists. The closest name we would be able to come up with would be the Intelligence, as you referred to in your first book.

I feel surprise from you. You are not aware that your book has a presence here? I thought maybe I was not supposed to reveal that to you, but I was just told by Sananda that it was okay. Your book was one of the sources I used to learn your language, even though I had been practicing before it was available. This craft has a special computer program that teaches all the languages that we will need to know, but I wanted something that was a little more what you call personal. That is why I had them program your book into the computer for me. After I got your English language down pat, I really enjoyed reading it. In fact, I have gone through it several times. With our system of learning it only takes me a few minutes to go completely through it. Even though our civilization is millions of years older than yours, and is so different from yours, I really learned some new things from it.

I mentioned earlier that I enjoyed listening in on the conversations between you and Creator. (Yes, I have about gotten used to calling Him Creator.) I understand the fact that He has just manifested a part of Himself so that He could more easily communicate and you could better understand Him and what He is conveying to you. Anyway, early on He said to you that He really enjoys using some of the unusual sayings that you Americans have come to use. They are not all included in the program; so I watched for them and have come to recognize them.

Then I was told that you had used only a small amount of them, as has our Creator. I am anxious to hear more of them. Once I learned to speak your language it was easy for me, even though we do not have a verbal language as you ones on Earth do. I am learning many other languages, and I have already mastered several of them, such as Chinese, Japanese, Russian, French, Spanish and several different dialects of the Middle Eastern countries whose religion is Islam.

There are so many fascinating religions in your world, and I find them interesting yet so archaic at the same time. I have been able to form an opinion about them, and I am told that I can speak freely with you about anything I want to. I do understand the concept of being judgmental; so I do allow the fact that ones are free to worship as they please. But for the life of me (I believe this is one those sayings), I cannot understand how there can be so many different systems when there is only one universal being, intelligence, god or whatever you may want to call it. And then all the fighting and killing. I could go on and on for hours, but I would only be talking about things that you already know; so enough of that kind of conversation.

Nonáme's Introduction

There are a few things that I want to talk to you about before I tell you about my part in our Creator's plan. I see in your mind, Bobby, that you pretty well already know enough about our civilization that I needn't go into it very much. It's mainly that we get all our needs from the universe and spend most of our time learning about all the different realities of the universe. It is mind-boggling (another term I like). Even as long as we have been learning about the universe, we have hardly scratched the surface. Once you get your memory back you will understand what I am talking about.

Also you will know about our solar system, since you were part of its beginning. And that is why you have a small amount of remembering. Since I was, or am, aware of your part in our beginning, that is one of the reasons that I was asked to be a part of this wonderful occasion. I was so excited after it was explained to our collective consciousness. (Yes, we do have emotions. We have made an extreme effort to maintain many of the traits that are known in many parts of the galaxy.) Bobby, I see that there are certain specifics that you seem to be unable to bring through. Or you believe that you cannot. We will have to work on that. Do you understand what I am saying?

Since I have dropped down in dimensional levels I understand a little about the problems that you have been having in the third dimension. I have been told by the ones in the group that have been in physical bodies in the third dimension that I really don't have a clue what it would be like, and I don't think that I want to experience it. With my good what you call imagination, I have tried to imagine what it would be like. And to think that you have spent many physical lifetimes in that dimension!

When I was picked out of many and approached to be a part of this, I was almost in shock (another emotion) after it was placed in my awareness. Even with so much that we as a race have learned about the universe, we had only heard rumors about what was going on here. We have only explored certain parts of this galaxy, even though it is our home galaxy. We had already decided to explore this area, and it was on our schedule when we were contacted about my coming to this area to be part of the team. In all of our travels (not the way most travel the far distances of the universe), we have never come across a system as dense as yours. The reasons had to be explained to me. Even the planets that are still in the early stages of development are not this dense.

Now, about my reason for being on the team... The closest description of what I do is what you call on your planet a psychologist, and you are not all that happy with psychologists. But you would be blown away with our methods. We will discuss that at a later time, if there is any time. I will be much help to you in your dealings with the inhabitants that you will be dealing with. Our Creator has covered all the bases, and He feels that this will be a great help to you. I have been told a little about you; so I can sort of help you from over your shoulder, so to speak. Sananda says that I should probably close for now and that we will talk again soon. We don't often have to identify ourselves with names; so I have just told the others to choose one that they would like to call me. So I'm telling you that also. I'm looking forward to what name you will choose.

Second Session

We only got to communicate for a short while last time, and that was mainly to tell you a little of who I am. Since Sananda and our Creator also told you some things about me, we can dispense with that. Except that if you have particular questions you would like to ask pertaining to me, just go ahead and ask.

Now, I have already mentioned to you that the nearest thing to compare me to would be a psychologist. You might think that a civilization as old and advanced as ours would not need the help of a counselor (another good description), but we have many other responsibilities besides counseling. Every event in our lives is so organized and in such precise order that even a slight deviation can be traumatic. It would be like what you call the "domino effect," and it does occasionally happen. Then the one that caused this effect would have to be reprogrammed, as you call it. That may seem harsh to you since you are such a "loose" society, but that is the way it is and has been in our society for eons of time. Everyone accepts that as a way of existence.

Everyone on our planet is pretty well aware of what is going on between each other and anyone we come in contact with. Once that is learned, very early in life, we begin many years of an extensive program. We do not have a schooling system such as yours. It is more like the computer system I learned your language from. After this and many years of just tuning in to the collective consciousness, all beings know what they are meant to do. (We counselors are chosen because we have an extraordinary talent that very few have.) No one complains. Everyone is equal; so there are no higher quality jobs as you call them. Also there are no "trivial" jobs such as janitors, dishwashers, garbage collectors and such. All that is done by what you call robots. They do all the menial work as well as any work that is considered dangerous. I could go on and on but that is enough. I kind of got carried away.

We will now talk about how I am to be of service to you. I am told that I will make a brief appearance with all of the team, and that will probably be the extent of my appearances. I have been chosen to help you as you meet with certain ones. If you have any difficulty with any "client," I can step in, look deep inside that person, and tell you how to continue dealing with him. I will know in a short time all you will need to know. I understand that you will encounter many different types as you talk to the ones that you have to give an ultimatum to.

I am so excited about this. I know that we will encounter unusual situations; so I expect I will be exposed to many new emotions that I have not experienced before. I can do all this remotely; so I will not need to be with you. I expect that I would intimidate many.

Do you understand what I am to do for you?

Bobby: Yes.

And are you in any way offended by what I have been asked to do?

Bobby: No.

The other members of the team feel that I will be much help to you.

Bobby: I will be very appreciative of your help. I expect that this is going to be a rather "tough" assignment.

I am glad that you feel that way. I would hope that we can work together very effectively in our dealings with the bad guys. This should be a very interesting experience for me. It will also be very interesting taking this experience back to my civilization and trying to explain to my friends and fellow counselors what this is all about.

Third Session

I am excited with the fact that I will soon come face to face with you and experience a lower dimension. I have been warned by the ones that have experienced it that it is really nothing to get excited about. I am still looking forward to it because I have always enjoyed experiencing something new; it's just in my nature. In looking at your thoughts, I see that you are the same in that respect. Like "seeing what lies over the next mountain" – a good analogy. But mostly I am looking forward to meeting you.

I didn't explain to you before that we have a different way of observing everything we come in contact with. It is similar to what you call sonar, that I understand is what your little creatures that you call bats use. However, it is much more sophisticated. We see an image and then that image is changed inside to enable us to "see" everything we come in contact with. We do not need light to see these images. It would be difficult to explain how we manage to be aware of distant objects, even stars. I will say that it is connected with thought transfer, the ability to just be aware.

I have seen that the Creator has been delving into the finer aspects of what you will be dealing with when you come aboard and then later when you make contact with the various groups on the surface. I have been spending much time in looking at these different ones and with their agenda. I am amazed at what I have seen. What I am now doing, at the suggestion of Creator, will help you to be more aware of what to expect. Creator, of course, could do this Himself, but He sees that as a conflict, a possible infringing on those beings' free will. He does not want any possibility of that happening. Since I began this I can now understand why that decision was made.

This elite group is the one that you need to be most concerned with. They are somewhat aware of what is to soon come about, and as Creator has said, they are preparing for any possible scenario. They are a very dangerous group. They do not intend to let anything stand in their way. This is why Creator recommended that I observe them very closely, since He is not quite ready to make you aware of their total agenda.

There are certain ones that are able to know when they are being "scanned", and they also have the electronic means of knowing this. They are now able to block most of their thoughts. It took me a while to be able to penetrate their shield and to be able to know what they are doing without them knowing. I am being careful in looking in on those ones' thoughts. They did not expect that anyone would be able to penetrate. (Sort of like your being able to penetrate Lucifer's shield from the inside.) Now, they are somewhat aware that I have been attempting, but they do not yet realize that I have been successful. They eventually will know but by then it won't matter.

What I have learned is that their plans are quite ingenious. I can see why Creator has asked for my presence at this time and has asked me to make you aware of their overall plans. You would have a more difficult time with them if you were not aware of their agenda. You would eventually succeed, but it would be time consuming, and I understand you do not have an abundant supply of time. Even though I know their plans, Creator has asked me to withhold it for the time being. He feels that it would be too dangerous for you to know at this time.

Our civilization is very well versed in the concept of how the universe works, the possibility of "good and evil", the dark side, and the fact of the positive and negative aspects of creation. But the concept of one being representing the "dark side of creation" was a little difficult to grasp at first until I looked at the whole picture. We were not aware of the one you call Lucifer. Now we understand that he is sort of a local phenomenon.

I am now getting well acquainted with this one called Lucifer, a very interesting being. He is difficult to try to scan because he is also able to block most intrusions. But I am still able to understand much of his way of reasoning as to what he is doing. I am well versed on what Creator expected of him in the beginning, and Lucifer agreed to all the terms of what was asked of him. But he has pretty well abandoned the conditions that were given to him.

In one sense Lucifer has great expectations about what he is doing, and then he is frantic in his efforts to complete what he is allowed to do before his time runs out. So that is really motivating him. He really believes that he will succeed in his efforts. He has completely blocked out any thoughts of failure. It will be very interesting when the two of you meet, and even more so when he is confronted by Creator.

Now that I have looked at where this whole concept is going, I am somewhat overwhelmed. The idea that it will eventually spread over the entire universe is mind-boggling. I keep saying this but I feel in my emotions something I have never felt before: this overpowering sensation of what you term excitement, and more.

More About Nonáme from Sananda

The one that you have called Nonáme has actually been here for quite some time. She was chosen early on by Creator but did not arrive here as early as most of the others did. She was, as you say, "tied up" with other responsibilities on her own planet, and her responsibilities were so great that it took some doing to be released from them for a time. If it had been for any other reason, it would have been impossible. She is a high ranking member of the Council of Governing body, actually the highest ranking member. It is a lifetime agreement once you are chosen. It took a considerable amount of time to find an appropriate replacement for her and then to work with that one for some time before she could be excused to do this honor for the Creator.

Nonáme is a real blessing for our team. Being of high caliber in her profession, she sees things, as she is trained to do, that we often overlook. She will be an invaluable asset to you when you are presented to the leaders of Earth shortly. She has been made aware of the major players in the game and has been studying them for some time now. So she will be totally aware of their agenda when the time comes. She is from a no-nonsense type of race but is still very sensitive to others and is a very extraordinary person. You will be amazed.

This is not taking anything away from your responsibility, because you will feel overwhelmed by what is going to happen. Creator is covering all the bases by adding Nonáme to the team. He is giving you all the help that He feels you will need. He feels that there will be much more resistance, and therefore danger, than was expected. Lucifer has been quite adept at what he is doing. He is going to play the game, as you say, "down and dirty." He is giving it his all.

17

Elijah (The Old Testament Prophet)

First Session

I have been asked to come to you for a period of time. I noticed while being aware of your current reading material that my name was mentioned. So it seems that the timing is right. I was contacted some time ago by Sananda and also our Creator to be present aboard their craft to observe for a while as they communicated with you. They wanted me to kind of watch and be aware of your mannerisms and all the little things that you do so that I could communicate with you on your level and both of us would be comfortable.

First, I want to be sure that you are willing to start a series of communications with another one of the infinite One, since you are in contact with several of our Creator's subjects. Very good. I do not want to impose on you if you are not absolutely willing. I will be here for a while to observe with many others what is about to happen. And if it is still all right with you, I will continue communicating with you. This is my first time to communicate like this with someone in the physical. I am pleased with the way that it is going. So I am looking forward to continuing this with you. At this time I have no plans to communicate with others, but then that could change.

I am pleased that I was advised to observe for a while and to get used to people and learn the different ways that people talk and react to one another. There are so many ways of expressing oneself that I am amazed. I can see how difficult it would have been if I had tried to express myself in the language of the past, even after a translation. I can use your English language due to the computer program that is available on Sananda's craft, which brings me back to my story. I am one of the few here that have the ability to speak, as many of the ones here do not speak vocally but communicate with thought transference. That is actually the best way of communicating because you can express your thoughts and emotions to a much greater degree. I really like some of your words – the sound of them when they come into my consciousness. There I go again. Oh well, we will eventually get to where I am going.)

I want to make you aware of who I am and a little about me. I have not been in the physical on Earth for more than a thousand years. As for so much of the information in your so-called Bible – "the absolute word of God" as so many have been saying for many centuries – there is really not very much truth to be found there. The "book of God", written by the "hand of God", is mostly the work of man. I see that you are a little concerned that I am merely expressing your thoughts because you have had those feelings also for a good part of this incarnation.

But believe me when I say this: even though those are your feelings and I honor those feelings, what I am about to tell you are my own beliefs. Those beliefs are from many lifetimes on the planet Earth and many episodes of the times between those lifetimes on Earth with many beings of higher dimensions. These beings have been on Earth and observed all of the facts and all of the untruths that have been recorded. So that's why I say that much of what has been recorded as history is not factual in many cases. That is now causing much turmoil in the beliefs of many of the religions that have sprung up on the planet during the past two thousand years.

I am aware of your first book and its purpose: to bring some truth and to expose the many untruths that have been propagated among most of the Earth's inhabitants. I am also aware that it is only to be a short and quick read

for the mostly younger generation that has not formed such a strong opinion that they cannot see the obvious truth. I believe that it is accurate as far as it is able to go in such a short version, and it will be very effective when it is re-released. Now that that is said I will get back to my point.

There is no point in elaborating on my life as a prophet in the times recorded in what's known as the Old Testament. I know that there is much controversy about the so-called "God of the Jews," Yahweh or Jehovah, as well as the many so-called "gods" in different times. But there was a time, as you are aware and expressed in your book, when our "Creator", the Prime Intelligence, the One and Only was expressing Himself to certain ones.

It was very difficult for us in those days of ignorance when He tried to explain to us what was to happen many centuries in the future. But of course, it was a seed planted to grow and be cultivated during the succeeding years of our existence and other existences we would have. So there were some truths that were kept and recorded.

When I tried to explain what was happening to me when I was taken up into a spacecraft, I hardly knew what to record. There are many now who understand what was transpiring, but it is amazing how many there are who still do not get the meaning – that there were ones here at that time in spacecraft from other dimensions. It would be difficult to explain how I felt when I was taken aboard, but it would not add much to what I want to say anyway.

Life on the planet at that time was very simple and pleasant at times. People just lived from day to day, doing the only thing they knew and that had been handed down through the generations. Nothing evolved very fast at that time. The energy could best be described as slow. Evolution was moving at a snail's pace. The people of that time only knew what their fathers knew, and they lived with very little thought to the future. But there were also times of war – the invading tribes of other nations and times when the Jews had their own wars. Much has been exaggerated, but there were still many wars and much destruction and bloodshed.

Enough about the time known as the past. I just wanted to give you a brief rundown on my life. The life that I lived as a prophet is only one very short episode of who I am. I am a part of the All That Is, just like you and Sananda and the others. I am living the lives and doing what my Creator requests of me. And now that request is to join you ones for a time. We will see where that leads me. I am looking forward to more sessions with you, and we can discuss why I was asked to join your team.

It is so exciting. And so are the times. What a time to be observing what is about to happen. I have been in several other areas of the galaxy, and the word has really spread as to what is about to take place on Earth. It has been known now for several hundred of your years. Along with the knowing, there is much excitement about the event. I am amazed at how many systems are represented here from so many parts of the galaxy and even beyond.

Second Session

It is good to be in communication with you again. Since I was not really up to date on what was happening before, I should be better informed this time. I have been almost, like your term, "force fed." (I have been learning all of your little terms, and I think I will enjoy using them!) And of course, that was my first time to do what you call channeling, or maybe automatic writing is a better phrase. Anyhow, I am now more acclimated to the situation at hand.

I am truly amazed at how critical things are at this time. I have not been in a situation where I would have been kept up to date on these matters; so I had to take a quick course on the events. The facts about what is going on and what is about to happen are very wide-spread, but only the general knowledge is known.

Now, I really did not know why our Creator had called upon me at this critical time. And I am still not completely sure that I will be of any use at this time. I was made aware of the fact that the three of us – you and I and Sananda – will make an appearance together when the time comes for the entire group to land on the surface. That should be an eye-opening experience for the inhabitants of the planet. I am certainly looking forward to it.

As was explained to me, I will also accompany you on some of your scheduled appearances with certain leaders of government, industry, and religion, as well as the Elite in the hierarchy – the ones that have gained control of most of the goings-on on the entire planet. Sananda and the others seem to think that my presence with you will be more dramatic than your presence alone. Then there will be times, I understand, when Sananda will join with us. I can believe that there would be some individuals that might even request his presence.

Bobby, I am really not up to doing much communication with you yet. I am still amazed that I was asked to help and even more amazed at what is in the future for the planet. When I was in a physical body and was given to see what was in store for the people of Earth, I really didn't know about all of this. I was shown some things that were to take place in these times, but it was really hard for ones in those days to fathom what was to happen and so alien to me. Even most of the ones that were prophets in days past were only somewhat aware at the time of the truth about the Earth.

I have been made aware of Gaia as the living, loving being that is the consciousness of this planet. It is quite fascinating to know more about the Earth and its future as a loving being in the universe.

Third Session

Although I have been talking with some of the other members of our Creator's assembled team, I seem to be the one who is not completely sure of why I was asked to attend this grand event. Maybe it is just that I had been resisting to some extent. I have been spending more time with the computer program for language, and I now feel comfortable with many of those spoken on Earth. It's amazing how many there are. I wanted to be able to speak and understand as many as I could before joining you and Sananda when you begin your dealings with your adversaries on the surface.

Creator and Sananda seem to feel that my added presence will be an impressive way to begin the talks that will start soon. And now I do want to be part of what is coming. This should be an exciting time. I hear it is appropriate to feel excitement for this special occasion, even though it will be a difficult time for most of the world's population.

Even with all the discussion and prophecies concerning those times, some of which I was involved in, I never expected to be in this time frame and be part of this event. For one thing, we thought it would take place many centuries ago. But I am pleased to be joining you back on the surface again, even if it is a third dimensional experience for a time. I can see that Joshua's assistance during all the turmoil will be of great value; I hope I can be just as helpful. The general consensus aboard this craft is that we will get started soon on this adventure. Everyone is abuzz with expectation and last-minute preparation.

More About Elijah from Sananda

We were all mildly surprised when Elijah made his late appearance, and it was an even bigger surprise for you. It wasn't that he was necessarily the last choice; it was that Creator had declared that he did not need to be here until the event was almost ready to take place. We have talked some and he is not altogether sure why he is here. He is still sort of plugged into the computer program. After he is made aware of the "modern" way of things, he will spend some time with the team on a daily basis to learn what they are doing; then he will spend time with you. That will only take a couple more days.

You are curious about my relationship with Elijah, aren't you? I figured as much. We were, of course, very much aware of each other, and it has been said that I was an incarnation of him. That is kind of a difficult question to answer. You are aware that the possibility of multiple personalities in separate lifetimes is much different from what is normally thought of by ones in body now on the planet. I'm just going to stay with what I said for the time being. I'm sure that he will be explaining more to you as he becomes more aware of why he is here.

Elijah will be making an appearance with you in the third dimension. He is very dynamic, as you are, and Creator has decided that the appearance of the two of you will be very explosive. Having the two of you together is going to be quite spectacular. Also, I will be with you for a time before I have to get back on board the craft.

The Transformation of a Planet

The Transformation of a Planet

18

Sanami's Introduction

First Session

Bobby, I am the one that talked to you earlier this day about having some time with you. I am a new energy that has only been in this area of the galaxy for a short time. Creator asked me to be here now for a very specific reason: to convey a message of utmost importance to you.

But first I would like to introduce myself and tell you a little about myself. The others of your team said they would introduce me to you, but I asked them to let me do my own introduction, which they have allowed. I am not familiar with any spoken – or verbal, as you call them – languages. As is the case with a majority of the universe, we only use thought transfer to converse with one another. I also do not have a name as you ones of Earth.

I am aware of the other new ones that introduced themselves to you recently. I spent time on the computer program, as they did, so that I could give you this message. It was a little difficult for me because I have no digits with which to operate a computer. But I was able to connect my mind, my consciousness if you will, to the computer. Then it was easy for me to learn all your Earth languages in a relatively short time.

I am able to understand any of the group, even if they talk to me verbally...in any of your languages. I just translate their sound patterns into thought patterns that I can understand. Even though some of the more highly evolved civilizations consider communication by sound and vocal words to be very primitive, I am fascinated by it, and with the many languages and dialects that you ones of Earth use.

I am from a far distant unnamed galaxy, or I should say unnumbered since that is the way your astronomers identify them. Our civilization is much different from yours. For instance, we do not have separate genders as you ones on this planet. We evolved from that many millions of years ago. Our existence is so far from what you could comprehend that it will be difficult for me to explain.

I have connected into your consciousness and am aware of your thought patterns. I was told that would be all right because I have contacted your other self and received permission to do that. I also got approval from Creator. I see that you are able to pick up on what I am about to relate to you, or at least a small amount. I understand that is not a normal condition for most of the ones on your planet. And that leads me to what I came to tell you.

Our civilization is and has been aware of what you call your Creator since our far distant past – longer than our records extend back. We were one of the first civilizations in this universe to develop a means of communication. We know there are multiple universes. We are also aware of the boundaries of the universe that you and we reside in, as well as the endless void that lies outside the seven universes.

We are beginning to understand that there may be other clusters of universes in that unimaginable void, similar to the clusters of galaxies that exist in our own universe. It is not that we know every solar system in every galaxy in this universe, but when Creator came to me and my civilization and mentioned your solar system, we instantly knew the location and all the pertinent facts about your system.

Sanami's Introduction

Since you have already sensed some of what I was going to say, it may not surprise you that I recognized that you were one of the creators of our system. But it was a surprise to us. We had always assumed that there was only one Creator, and when He made His presence known to us for the first time we were overwhelmed. We never expected that to happen. We do not have legends about what you call God ever being on our planet. In our knowledge of many, many planetary systems in many galaxies, we are not aware of any that have a legend like you do that God would manifest as a human being to save you from yourselves. I am told I should not take that any further.

We as a civilization are extremely happy (yes, we experience emotions) to have become aware of your whereabouts. We are extending a special invitation for you to visit us as an envoy from your planet. Since there exists no world council for Earth, and no one to designate as ambassador, Creator said He would designate you since you have been such before. You do not need to decide until we are able to meet in person.

Our Creator – I am speaking of the Prime Creator – manifesting Himself as a consciousness, asked if I would transport myself to your solar system for a specific purpose. After He explained why, I was not sure it would be possible. But He made it clear that He would assist me and make it possible. (Our civilization is not the only one that travels from point to point in the universe instantaneously, but we are the first to accomplish that feat.) As soon as I agreed to do what He asked, I was instantly here in your solar system. Our Creator cleared the way for me to appear in your small spacecraft.

We are unique in many ways. One thing we can do is search each being's consciousness to discover what their personal choice in a friend or companion would be; then we project ourselves as that. That is not a false presentation because we actually become that image and retain it for as long as we exist. We become that personality and never take advantage of that, no matter with the circumstances. I have already made that connection with you ahead of the time that you will bring your physical body aboard this craft.

Creator has asked me to spend some time with you when you come aboard. I told Him I could come into your presence while you are still on the surface, but He replied that would not be appropriate for several reasons. My presence on your craft has created a sensation among the entire crew, as well as your team. I have already spent time with the rest of your team, but Creator said He wanted me to spend more time with you because you will be required to return to the surface for a time after being on board the craft.

I will not be able to fully explain why I am here until you are on board where we can react with each other in person. I will not replace any of the others that you have already talked with; I will enhance your presence when you meet with the various ones on the surface. I know what you are faced with and will be of much help in your face to face meetings.

We are a civilization with much honor; that's who we are. We could never break that trust; it would destroy us. The thought of being dishonorable has not happened in millions of years, other than to observe it in others. What I offer is very unique and unknown in the rest of universe. After Creator asked me to join your team, our council decided it was time to share our capability. I am anxious to meet with you.

More About Sanami from Creator

We will talk about the new energy that came to you last night. Some things were left unsaid that you should know. Their civilization has no gender as such. Their bodies do not contain both genders as some beings do. They do not split in two as some do. They have a unique way to expand their numbers. There are no surprise births. They control their population to the exact number they desire. That is one reason for them to be here at this time. They will not be here in great numbers. But eventually after the second shift – the consciousness shift – takes place, there will be a few dozen of them who come to this planet to help bring the remaining survivors into the reality of the higher planes.

Even though they have emotions, as was mentioned last evening, they are not emotionally attached to any new souls brought into existence. They have risen to the level where they can create new beings by thought. All new beings are identical when they are first created, and then they move into the area where they are needed. They are only brought forth from the universal consciousness when there is a need.

Occasionally they expand into a newly discovered planet. At that time they bring into existence as many new entities as necessary to develop the new world. They never use any of their existing population for that purpose. The new beings are programmed to do what is required to settle the new world.

You are thinking that they are almost robotic. And in one sense that is true. They have evolved over many millions of years to be where they are now. They bring forth from the universe everything they desire, which is what I intended. That knowledge is what they will bring to the beings that will inhabit the new Earth. It will take some time for this to be taught, but these ones have patience...and "time", which is the best way to put it for the time being. You will notice that I am, as you like to say, sparing no expense for the ones that are ready to accept My help. And there will be many. The rewards will be great.

The one that you met last night has developed to the point that there is absolutely no situation that could arise, no foreseen or unforeseen circumstance where they would not have an immediate solution. This will be very helpful for you and the others of the team. It will mean the most to you in the beginning because of your role here on the surface.

Sanami's Second Session

Since we had our first communication, I have been looking over your personal file. (It was given to me at my request; I found that it was open to all the members of the team.) It was intriguing going through it and finding out all about you. It was evident that you have been a busy being for most of your existence. I was impressed with your achievements. That certainly supports the invitation to come and visit my home planet.

We have actually almost rebuilt our planet because it has had beings on its surface continuously for an almost immeasurable period. And it is not our first home planet. We had to move en masse because it was time for a renewal cycle. That cycle was very different from what your Earth is going to experience. It was time for our planet to be absorbed back into the oneness of creation. We were already extremely advanced at that time; so we were aware that it was going to happen.

We found a new planet a few thousand light years away in our same galaxy. Our planet is incredibly beautiful to our way of thinking, but I'm afraid that you ones from planet Earth would not find it suitable to your needs. Since we are vastly different from your race of humans, we have completely different needs. Also, it is much larger than Earth, but still it has only one sun.

Neither heat nor cold has any effect on our physical bodies. We prefer that the temperature be in the range of 100 to 200 degrees Fahrenheit, but it is not critical. I find it reasonably comfortable in the space of your craft. I am being told that on your spacecraft you have to have a variety of environments to suit the various species of conscious beings on board. You will be shown this when you come aboard. I got carried away talking about my planet, but talking about it gave me a chance to use your language.

When we visit other systems that we can communicate with, we quickly learn their language and method of communication. Our civilization was aware that there are many systems in the universe that use a vocal language like yours. I am fascinated by it even though I don't normally use such a system. I communicate with most of the ones on your craft, and I am having "a ball" using your language. It is very descriptive of all the things I have encountered so far. Our system is similar to the one used by Nonáme, but it is even more simplified. I will explain it to you when we can meet in person. We will find it relatively easy to communicate since you are able to pick up all of the ones that wish to talk to you.

I want to tell you what I have learned about how I will be of use to you and your team. I mentioned a few of my capabilities before, but since now I know a little more of what is expected of me, I can discuss it a little more intelligently. So that the ones of this planet can be sure that there is life in the entirety of the universe and know that they are not unique, our Creator wants beings from various places in the universe to participate. There will be a very good cross section of beings from many regions of the galaxy and beyond. So it should be a very interesting event, but probably a little traumatic for many on the planet. My appearance is not like yours.

I will be on board the craft with you when you make your first landing and reveal your new identity to the population. (I know it is not the first time you have made your presence known, but it will be a first for most of the ones here now.) After the meetings get underway, it will not be necessary for me to be with you in the physical. I can be of service to you from your spacecraft just as well.

My expertise will be slightly different from Nonáme's. Together we will be able to give you a complete picture of your adversaries' thoughts – past, present and future. We have been cautioned about free will; so we will be aware of that at all times. We will also be aware of judging others unless they are interfering with free will. I understand that Creator will be advising you too; so you will be quite busy sorting out all the information given to you. But it will be done in such a way that you will be aware of all of it.

Third Session

Since I last communicated with you I have been looking at all the various scenarios currently taking place on the surface of the Earth, and I am totally amazed at what is happening. It has been so far back in my planet's history since anything remotely comparable to this has occurred that I can hardly believe it could still happen on this third dimensional world. But I have come to understand that this planet's future impacts the entire universe. So it is important that this planet be protected. And it will be, regardless of what seems to be taking place. I understand that this has been your responsibility, and I know that there have been many close calls in Earth's long existence.

Now this phase – its tumultuous history – is just about over. Earth is being prepared for its awakening. These are exciting times; is it not so? Just to be invited here to be part of what is to come is quite an honor. Of course, since you have been here during most this planet's existence and have been involved in its evolution, you probably have a totally different feeling, I would expect.

I guess you are still wondering why I was asked by Creator to be part of this. I only gave you a brief overview earlier, and actually at that time I was not sure myself. I am just now beginning to fully understand the reasons for being here. Nonáme gave you a good understanding of her purpose: to make you more aware of the agenda of one of the groups, the most difficult group to deal with.

In looking at what you have written, I see that you have considered the possibility that other groups have had the desire at one time or another to take control of the planet. I see that Lucifer is partly responsible for that. My home planet, like Nonáme's, was unaware of his existence. However, we are aware of wars and have experienced them in our distant past. It seems that Lucifer only has to plant a seed; then he nourishes it with negative energy and watches it grow.

What an amazing way of controlling another. I see how so many could have given up their free will without even knowing it. I have reviewed Earth's different time periods, and I have observed that many groups, and even individuals, have had the desire to rule others and completely control their lives, and have wanted to control the entire planet. But no one has been able to do that, although some have come close. I am only observing at this time, partly for my own benefit, since it is appropriate.

It might not be so important for you, but what I want to focus on is the various religions. Even though the majority of these do not have the capability of being a real threat, you would do well to know what they are thinking. It is unbelievable how many of them truly believe they are the only ones living the way God expects them to. To them all the others are "wrong" and most think those will go to the place you of this planet have created known as Hell. And there they will be tortured for eternity. It is difficult for me to even comprehend such an absurd notion. I have been informed that normally it would not be acceptable for me to state such as I have, but now it is time to tell it like it is (a popular phrase, I understand).

I have observed that in the distant past some religious groups did have control over a large percentage of the so-called civilized world, and that still exists even today to some extent. They are the ones that believe only they know the truth, and they are not going to peacefully give up their beliefs. There are many that have such strong beliefs they will not even look at any other. When all these rigid beliefs are combined, a large percentage of the world's population is involved. They will not only be angry about what you are saying, they will be angry with each other.

Therein lies the problem for you, and even for the rest of the world. That is why I will be watching so closely and letting you know what I see. I understand about free will and that it will be a sensitive issue for you. Creator has been very specific about that. All you will be able to do is relay the truth that will be given to you by Creator. Then it will be up to them. Since I am mainly going to be helping you deal with religious groups, I have spent a considerable amount of time with Sananda. He shares my concerns about what you are up against and about what he will have to explain to a large percentage of the population.

There is another faction that we need to discuss because they are also extremely dangerous. Even though they are considered part of the elite corps, they have their own agenda. These are certain ones of the military/industrial complex plus heads of state as part of this group. These are the ones that possess secret technology, either given to them by some of the alien groups or reverse engineered from UFOs that have crashed or been shot down. This group has kept the UFO situation from the public for more than sixty years.

They have a somewhat different agenda from the rest of the elite corps – they are obsessed with war. They live it; they thrive on it. That does not necessarily mean they have to be fighting a war all the time. Just the idea of preparing for war is satisfying to them. The military/industrial complex wants to keep building more weapons and deadlier weapons of war. That means convincing the population that it is necessary so they have the means to keep it going. They must be able to appropriate the monies needed to keep the war machine alive. This group will also not give up easily.

Sanami's Introduction

So, my friend, do you see why it is imperative to repeat over and over the situation you are faced with? I am joining the others who have been telling you this, not to frighten you but to be sure you totally understand. However, we do not see that you will be in any life-threatening situations because of the amount of protection you will have.

The Transformation of a Planet

19

Joshua (From the Old Testament)

First Session

I am here at our Creator's suggestion to talk to you a bit about what you are about to face. I am a familiar being in your biblical lore, even though many of the things that have been written about me are not true. That has happened quite frequently, and there are many reasons for that. I just wanted you to know that. I am the one you have known as Joshua, the one who supposedly blew the walls down. That event is only partially true, but I am not here to discuss my life.

As with a great many of the characters in the Bible, we all had our share of adversaries; I did for sure. But it's too bad that more wasn't written about the good times that most of us had. It was not all battles, but I am aware that is what gets the attention of the ones that are studying the Bible or other writings. They are not interested in the day to day boring events of an individual or even in a nation's real history.

I am here because I have experienced some of what you are about to experience. But I did not have to come face to face with Lucifer, which is one of the events that will be required of you. Also, I understand that you have already experienced that in times past. I bring this up because I was certainly aware of his presence and what he is capable of.

Things were very different back then, as you know. We were aware of the presence of the god that is known as Jehovah, the one that some call the god of the Jews. It was quite a shock to find out that this being was actually Lucifer. He tried to control and succeeded in controlling the lives of the Jewish nation, and for a long period of time he was successful in keeping the nation totally in fear of him. So that was what we expected a god to be – one to fear and serve absolutely without question.

It was much later before I realized that was not the true God's way. God is not vengeful and does not control everything you do. At the time we did not understand the term free will, that you were allowed to do and be anything that you wanted, and that there was no judgment except by yourself. I mention this only because I can now observe from this side of the veil the true intent of Lucifer. He is carrying out his part in our Creator's overall plan for the new Earth and then on to the entire universe. But it seems to many that he has gotten carried away in his zeal in playing out his part of the game. He is certainly enjoying what he is doing but will probably have to eventually pay the price for what he is doing when his part is finished.

I am here, Bobby, to add my point of view to others you have been given because we are able to observe what Lucifer is planning to do in the near future. He is different and much more powerful now that he has managed to sow the seeds of hate, causing the people of Earth much fear. He has also caused the people to hate each other to the point of killing millions in wars, so-called ethnic cleansings, and the genocides of entire cultures.

I know that you have been told not to underestimate him, and I am also stressing the importance of that fact. Since I spent a good part of my life fighting battles to defend ourselves, many of which were not necessary, I am well aware of what you and the rest of the team are up against. The difference is that the weapons at your disposal are so much more sophisticated than mine. But then, you will only need one weapon, and that is the weapon of truth.

There is going to be a big surprise to many of the inhabitants of the planet, since they will expect Sananda (Jesus to them) to form an army and physically slay the enemy. Our Creator has said that you will not be able to keep the dark forces from completing the acts of terrorism they have planned. From what I have seen, these acts are horrific. And all you can do is watch and wait. I have dealt with many of the things that you will face: confronting the heads of state, the many religions, and perhaps the hardest – the military. I realize that in many countries that will mean dealing with only one entity, for this one controls them all.

That is what you and I have in common. Not all the battles in those days were fought on battlegrounds. It is not well-known that we tried to negotiate with the kings and leaders. In most cases this was only one man and in an isolated case, a woman. In a sense it was easier to deal with the supreme being in power and no others. But it was also difficult at times to deal with the one person that controlled everything in that country. In my country I was chosen to be that being. Diplomacy did not work well in those days; so we usually had to fight after all. We of the Jewish nation were totally unaware that Lucifer, whom we thought was our god Jehovah, thrived on the energy of war and the fear and hate that went with it.

So basically you are faced with the same situation. Once again Lucifer is controlling much of the planet, and there is no negotiating with him. He is single-minded and has only one agenda: to win this game. He is very aware that his time is running out; so he is pushing to the limit. He is the great deceiver; he will say anything that he feels will give him an advantage. But he will not honor anything he might say if it is not to his advantage.

I am feeling, though I have not been told this, that he may manifest a body and appear in the physical. To my knowledge, Lucifer has never done that before, but only the Creator has an indication of what he might do. That might be the showdown between the two of you that you have been told about, and it may happen in the physical. That could be an interesting situation, and if that is the case, you will have a very large audience rooting for you.

Again I say, be very alert and open to the unexpected. Naturally you have the big advantage, even if Lucifer doesn't think so, because you have our Creator in your corner giving you signals as to what to do and say. We all know the outcome, but getting there could be dangerous. It also might be fun. I hope that I have been of some benefit to you and that you will consider all I've said. I'm glad we've had this chance to communicate with each other. I expect that I will still be here when you are brought aboard this most incredible spacecraft, and I hope we will get to meet.

Second Session

It is my impression that I might be required to attend at least one of your meetings with the adversaries before many more days pass. I will be honored to lend my services to your encounters.

As I said before, I was shocked to find out that the one who identified himself as our God Jehovah was really Lucifer. However, when I look back over our dealings with him, it should have been obvious by what was happening between him and the Hebrew Nation. During those times, there was a rather large number of spiritual followers that were hungry for some kind of association with their God. There had not been anything like a personal God in a very long time. So the timing was perfect for Lucifer to make himself known as our God Jehovah, even though our true God was in our presence.

I vaguely remember my experiences with Lucifer. It has been quite some time since my contact with him. At that time, we as a nation did not look at Jehovah as the God of the Hebrew Nation; he was just God to us. He led us to believe that he would always lead us to victory. We had no other concept of God other than the fact that if we did not obey him we would suffer bitter consequences.

After my life ended, I spent time contemplating my life as a servant of the God Jehovah, and I knew I would have to make amends for that life. We reach a certain point after many lifetimes when we look at them as experiences, with no judgment of others, regardless of what transpired during those lives. Even so, I felt it was necessary that I return to Earth immediately to heal some of the wounds I inflicted on others, as well as those I received. But my guides suggested that I do the self-healing before I returned to Earth for a chance at going in another direction. So it was a long time before I returned.

I wanted to return to the same area and continue what I had been doing, but you can really never go back to the same situation. Besides, you usually forget what you had intended to do when you get back. It happened that way with me, although I did manage to work off some of the karma I had accumulated during the previous lifetime. As of now I have managed to work free of all the old "baggage" from my lives and am waiting to help others do what I have achieved. So I am offering my services to you in whatever way I am able to assist.

I am fascinated by what is to come. We on this side of the veil can see that the cleansing is a necessity even though we are going to have some rough times. We can see that only a small percentage of the world's population will make it through all the upheavals that lie ahead.

I have spent a little time with Narno, and I have been told about some of the experiences you two have had together. I feel that you and I are similar because we are both warrior types. We warrior types see things differently from others. That must be the reason Creator has requested my presence here at this time. I may not be much help during your negotiations and explanation of your purpose for making an appearance, but I definitely feel that I can assist greatly with all the chaos that will occur after the cleansing. Creator has hinted that will be the case. So I am looking forward to being under your command as you begin to bring order to the world once more. I will be able to serve the Creator in a real sense this time around.

I elected not to go through the birthing process for this experience because I know it is possible I might not awaken from the density of the third dimension. I was allowed to appear on Earth as a mature adult with full memory of who I am so that I would be ready to work for you in any way you see fit. Creator has indicated that it will be up to you as to how I may be of service. I await your response. I am anticipating the day when I will feel like a warrior again because it will be in a different capacity this time.

Third Session

The computer program for languages is a marvelous instrument. I just love the way that you can learn a language so quickly. We don't get to do much practice aboard ship because there are several beings that do not have the means for vocal communication. I know they say that we ones who still use that method of communication are rather archaic, but sometimes there are certain words that have a feel about them when they sort of roll off your tongue. I miss that, even though thought transference does have its advantages.

Very well, my friend, I am ready in more ways than one and anxious to get started on this once-in-a-lifetime, or once-in-an-existence event. I might have been a little premature in asking you if I could be of assistance, but I am eager to be back on Earth's surface and make myself useful. I believe that I did get an approval from you to join you when the time is right.

I really feel that I can be of help. Maybe not in the negotiations with the various groups on the planet, but I feel I will be better suited after the talks have all been completed and maybe even before the shift occurs. I expect there will be much chaos when the ones on the surface have to face what is to come. I am not totally sure of what will need to be done at that time, and I have been very observant of what Creator has been telling you about free will.

The concept of free will was an unknown to us as a nation; so it will be a new experience for me. It is amazing that something as important as free will for all souls was so little known and even less understood. But I have been paying close attention to what Creator said to you, how insistent He was that free will is absolutely important, and how we will have to be extremely careful not to infringe on the free will of anyone, especially the adversaries. I have it so ingrained in my consciousness that it will now be second nature to me.

Even though I was not aboard this craft when you were asked to begin corresponding with many of the team and others, I have gone back into the records and looked at all the writings you have done, so that I would know all that was given to you to record in the physical plane. Some incredible knowledge has been given to you. I see now what the other team members have been talking about – about all you have been through these past few years. And I understand it was a necessary experience for you, you alone, to go through. Is it appropriate for me to discuss this with you? I see I should have asked you first.

Bobby: No problem. I am having a little anxiety over the fact that everyone is, or soon will be, an open book. But there are no secrets; so yes, it is okay.

That is good, but I will ask the next time as a common courtesy. There is so much to learn, and that is why I went back over the records of your writings. It is so incredible. It has not been mentioned and I am curious: why are you not already being called Michael?

Bobby: It is just that until I leave the third dimension I am still Bobby.

Maybe it isn't appropriate, but I have to say this: I really have a strong feeling about you. I am going to look in the records and find out what that is all about. I am anxious to know if my feelings are on track.

Bobby: I have also had some feelings – maybe not as strong, but they are there. It is a little difficult to remember in this density.

Yes, I can see that it would be. If you like I will tell you what I find. I do enjoy talking with you, Bobby, and I am really anxious to meet you face to face. And from what everyone is feeling, it probably won't be too long before that happens.

More about the Team from Creator

We have now covered all the main players. The remainder of the team is not any less important; all are necessary to My overall plan. You have been told by Sanat and St. Germain their reasons for accepting My request; so you are aware of their capabilities. You are familiar with some of the rest of the team, but there are some you are not familiar with. Their jobs are mostly behind the scenes. They will constantly monitor your every move, but no electronic gadgets will be required for that purpose.

I chose them for their extraordinary abilities with their minds – they will be aware of your every thought as well as the thoughts of the ones you will be dealing with. You are in the hands of the best. Each one has his own expertise, but by linking their consciousnesses together they will be absolutely phenomenal. Nothing will escape their observation of your being. However, if they are not needed at a certain time, you can stop their surveillance by indicating that.

Essentially, I wanted to give you what you need to know about the ones that you will be in close contact with during the entire process. As you know, Narno will be with you at all times. Sanat and St. Germain have a somewhat different purpose. They will receive the information gathered by the others, interpret it and relay the information to you as needed.

They will have a monitor that will allow them to observe you continually via video and audio, as well as to record all that transpires between you and the ones you will be meeting with. One of their main functions is to create a permanent record. Everything will also be entered into what you call the akashic records. Nothing will be left to chance. If anyone or any group denies something they said or did, we will be able to confront them with the evidence.

It will be a team effort with you at the core. Believe Me when I say that you will very quickly adapt to being at center stage. We will do some dry runs of a sort; they will be done by computer. The team can program any conceivable scenario. It's going to be quite a show by the time it is completed. And yes, we will do a lot of filming and recording of the shift itself, although it will be different from the way it is done by you ones of Earth.

Do not be concerned that there will be nothing for you to do if I keep bringing forth surprise assistance. You will still be in charge, calling the shots. It's just that I want to have an unlimited amount of ammunition at your disposal. They will all be operating from higher levels to send you whatever you may need. And only when you call for it. That includes most of My help also.

There will be times when one or more of the team will be with you in the physical when their presence is needed. And there may be times when you call for their presence. All of this will be at your disposal, but it will be your responsibility to decide how it is used. So now you can see that your role is very important.

20

Other Relevant Sessions

A Word about Channeling (Gaia)

For most of humanity's existence you have been fed much disinformation. Now that does not only include the information coming from the ones who are mostly in control, but also from the various beings in the invisible realms – the higher dimensions. Many of the ones from these realms like to play games (actually, it is all a game as you are well aware), but the ones in the lower dimensions (fourth) are many times not as informed as many of you ones still in the physical. I know that you are aware of that, but most of those still in the physical are not. So the ones that have agreed to be a channel for some unseen being are not always given accurate information. The ones doing the channeling and the ones that are listening to or reading what is coming from the supposed higher realms need to be aware of this and use discernment, so as not to be tricked.

There has been much good and needed information that has come from channels such as you and others. Also, much has been given, especially to you, Bobby, that may not be appropriate for that certain time frame but will be pertinent at another time. We, as well as other ones from the higher realms, sort of play games with you as well. But our reasoning is to get the ones that are receiving information to determine for themselves whether what has been given is appropriate for them or is total disinformation. You are quite aware of that, for it has caused you much discomfort over the past several months. But you are now beginning to understand the reason for this, and as we get farther into the time dispensation, you will come to realize that everything that has been given you is for a reason.

Bobby's Question for Creator

The Author:

I had asked how the One that was communicating with me was able to have a dialog with physical beings on the surface when the One we have named God is not a being as such, but the totality of All That Is. Therefore, there would be no body, either etheric or physical to be able to have a conversation with physical beings.

Creator:

I simply created Myself as a conscious being just like all the other beings that I created, but with one big exception. Even though I created all conscious beings in My likeness, with the full power to create, each is still only a tiny fraction of the totality of My being. Whereas I gave Myself, actually many more of My many selves, all the powers of creation I have. To many of these, I shall call them Subordinate Selves, I gave a specific program for a specific reason.

And I can create an unlimited number of selves as needed. The only limitation is that they cannot descend into a physical body. The created Self that is presenting this knowledge to you was created specifically for the purpose of seeing that the Earth completes its ascension into a higher dimension.

Another Subordinate Self (Aspect of Creator)

So, you are the one taking down all the information given by various ones, mostly from the team that was assembled by our Creator. I am not a member of the team, but I have been aware of it for some time now. I understand that you do not know the names of and have not been personally introduced to several of the team members. Well, I will identify myself shortly.

I am currently, shall we say, stationed on the planet Earth, and I have been here for several years. I also came here at the request of our Creator to monitor certain things. I have been in this same body for all those years. I can come and go between the physical and the higher planes as needed, but I do not come into direct contact with humans very often. When I do, it is mostly with the common everyday beings that do not have anything to do with the hierarchy, the governments, or the leaders of any of Earth's numerous religions.

I am here to keep an eye on and monitor the pulse of the entire population of the Earth. I can do that more accurately with the common people. Before I came to the Earth's surface, I learned most of the Earth's various languages and dialects so I could converse with almost any group on the planet. I do not communicate directly with some of the smaller indigenous tribes, but I use the universal language of thoughts. I do very well at understanding them and convey My intent to those ones as well.

You might be surprised at the fact that I can get a clearer picture of the true goings-on from those that you Earth beings call primitive. (You and some others of the planet are aware that they are really the very special and highly spiritual heart of the planet.) In most cases they are in contact with Gaia, the consciousness of the planet. They are also able to read the thought waves that are sent out by most of the inhabitants of the planet. So I can get a very good picture of what is taking place at all times.

That does not mean that I don't make contact with others; I sort of compare the energies that I get from the indigenous tribes to that of the so-called civilized tribes on the rest of the planet. I have come to the decision that the reality of who the really civilized ones are is reversed! The indigenous tribes often have strange beliefs and customs, but they do truly believe in their traditions, and they are life-giving and life sustaining to them.

I have just been made aware that you have similar beliefs about those so-called savage tribes and have much empathy for them. That is good. It is not very often that I pick up that kind of energy from the ones that I monitor. I have not read your energy that completely yet; I wanted to do that as we go along. I have been made aware of what your purpose is in the very near future, and I must tell you that I am not sure that I would want to be in your shoes. You have a tough job ahead of you.

My job is almost complete, and by the time you ones make your landing I will be finished. I was sent here with a very specific purpose in mind. My report to the ones who sent Me is for the most part for Me to give My opinion of what should happen in the coming days pertaining to the cleansing of the Earth's interior as well as the atmosphere surrounding the Earth.

Maybe I should go ahead and identify Myself now. I am another Subordinate Self of the One. Each one that has been sent here is here for a specific reason. We have only made this fact known to a very specific group, and there is no one else on the planet that is aware of this. I have purposely not been made aware of your presence until very recently. And as I said earlier, I have not monitored your energy before. I wanted to do that as we converse with each other.

I do not have a name, just a certain frequency. There are other Subordinate Selves here also. What happens is that the One separates Himself into different aspects of Himself. (I use the masculine since that is basically how He is identified.) I could even say how We are identified since I am a part of the One in a special way. We are different from you and the others in that We have the full acknowledged powers that the One has. Even though any time that the One separates one of us from Him, We are not really separate from Him. I say that so that you can understand that each Subordinate Self that is released has one specific job to do, even though We are all in contact with each other at all times. Our awareness is as One because We are of the One, but as a different sort of creation.

Actually, this is the first time that He has divided Himself in such a manner. (This has not been revealed to anyone outside of this planetary system before.) Because of the situation here and the importance of this project, He decided to do this. I hope you understand. I should also tell you this: the part of the One that you have been communicating with, the first of the Subordinate Selves, is, as you would say, in charge of the operation. When I am finished, I will just merge back into the Oneness as if My particular energy never existed. And what that means also is that the information and decisions will be absorbed back into the Oneness also. But the records are there for the ones that know how to access them. It seems complicated, but it is really very simple.

This sort of gives you a completely different feel for things, does it not? You are being given many new aspects of what creation is and how it works. Even as an "old soul", as you are called on Earth, you are still in learning mode. That is a constant. You are aware of some of this in your true form, but you must know that most of what I told you only took place after you dropped down into the third dimension; so you could not have been aware. That should explain the situation to you so that you can now better understand what I am telling you.

As I said earlier, I am almost through with what I was sent here to do (monitor the consciousness of Earth's population). When I am absorbed back into the Oneness, the final decision will be made concerning the severity of Earth's shift. I know that you have been told that the decision was already made, but you were just sort of being stalled until now. There will not be a waiting period, no subcommittees; it will be an instant decision that will be forthcoming very soon. I could be absorbed at any moment. I will not know when it is going to happen. I have been given the authority to reveal to you ones what I will recommend.

Because of the way that the Earth has been treated, and because of the fear and anger and other negative emotions, the planetary consciousness is not improving, contrary to what many of the ones that are channeling or receiving the information are saying. They are afraid that if they tell of the other reality, they will have created the inevitable. The shift has to happen; it is out of your hands. The report that is My opinion on what is to be done is only a recommendation on the severity of the catastrophe that is coming. I say "catastrophe" because that is the overall feeling of the planetary consciousness.

Gaia has been made aware of My recommendation already, and that is that the shift needs to be made immediately to stop the utter chaos and destruction that is going on planet-wide. On a scale of 1 to 10, 1 being the least amount of cleansing, I am recommending a 7. I know that seems severe, but just be aware that a few years ago I was prepared to recommend that it be a 10 – total destruction of all life forms and a completely new beginning. But because of certain requests, yours being one of them (and not because the inhabitants' consciousness has improved), I changed My recommendation to give the people of the Earth another chance. But it needs to be done immediately. Without delay.

A Few Thoughts on Religion (Creator)

There have been ones that were sent here over the past centuries to reintroduce to the planet the kind of love that was here many millennia ago in the Earth's ancient past history. And you know the results of that. They were aware of who they were and why they were here. They had a following for a time. The last being was My son, your brother, the one they called Jesus who now calls himself Sananda.

232

Now, all of these ones were aware of what would happen to them when they began giving My message to mankind. While they were on Earth, they were not believed; they were ridiculed; they were even killed. Then at a later time, after the message was lost or forgotten, the followers began to worship the messenger, especially Jesus, with the false belief that the only way to Me was through him. Supposedly, he was the only path to eternal life with Me.

There are many on this planet who are preaching love and the Golden Rule. But to some that means, "Do unto others as they do unto you." Some say there will be a great turnaround when all will come back to Me and then there will be heaven on Earth for all believers. But I see inside their souls, for that is where I reside. I do not see compassion. I see that most people do not even believe what they are saying. So how can Universal Consciousness respond to their prayers if they are not being honest with themselves and do not really believe what they are saying?

There are some who have faith and believe what they say, but their belief is weak. They know what they want but do not have a strong enough belief to make it happen. Because thought forms are real they can make it happen, but it takes focus and a strong desire for it to happen. If ones in the third dimension could see what those in the higher realms see, they would be able to understand what I am about to say. However, since they cannot see from the higher viewpoint, they may feel angry and frustrated. But this is the truth.

All the inhabitants of this planet possess free will; so they are allowed all things without My judgment or condemnation. (But not all understand that they will have to answer for any and all of their transgressions – the law of karma, as you of Earth have termed it.) This especially pertains to the right to worship as they please, or not to worship at all. Over the millennia this has led to countless religions, sects and cults.

233

An endless stream of believers has been drawn to a particular individual who has convinced a few souls that he has been given the ultimate truth from God. The individual has sometimes been able to draw believers with some sort of show of so-called miracles. Since all of My created beings have a built in awareness of Me, it is only natural that they would follow the urge to find Me. Once an individual finds that he can prey on that urge, they are halfway there – another cult is born. Now we are only talking about cults here, and most of the population is wary of these.

But whether they are cults or churches, what is the difference? Mostly it is a matter of how many people are drawn to them. If a religion is not based on universal truth, then it is still what I would term a cult. And here is where it is going to get somewhat nasty for the both of us. As I have strongly stated many times and you have written, because of free will people are free to follow whomever they choose, even Lucifer. Again, many at this time are following Lucifer knowingly, but most are following him unknowingly.

Now, I am not saying that all the world's religions and sects are wrong and are leading their followers into hell and damnation. (That would not be possible since hell and damnation do not exist except in people's minds. You have been browbeaten with those two words long enough.) But when people choose to follow certain ones, sometimes into the unknown, they need to understand that they are often being led away from finding Me, assuming that is what they are searching for. I am the easiest Being in the whole of creation to find; yet most seekers spend an entire lifetime looking for Me without finding Me. I would say that it seems they must be looking in the wrong place.

Most seekers feel they must go to the leaders of the church or other institutions for a path that leads to Me. "Which way do I go?" they ask. The leaders will usually give you an answer, but each place you go will give you a different answer. In a sense that is okay, because I am everywhere; I am all things. *Any path will do.* How can you *not* find Me? You could look under a rock, behind a tree, in the sky or in the waters of the world's oceans – I would be there. But if you do not know what you are looking for, how can you find Me?

Since the building of the first church, from the most primitive altar to the grandest temple, all worshippers have been told that is where you will find Me. But I tell you this: you are just as likely to find Me under the rock or behind the tree. If I had My choice – and I do – I would rather be found in nature than in the stuffy religious buildings of the world. They have been built by humankind for humankind to worship Me. I have no objection to that. But what about the "savages", as you call them? What about those who live deep in the rainforest or the deserts or in the ice igloos of the far north? You will find Me there too.

If you have dogma that must be followed and rituals and tradition that can't be broken, then it seems to Me that you do not need Me there. Maybe you don't even want Me there. It is all right to go to buildings to worship, but what I am strongly suggesting is that it is absolutely unnecessary. I am also with the believers who do not have churches to go to. Think of what could be done with the money that is spent on fine temples if it were channeled where it is needed.

I am the easiest of all to find, for I am part of you; I dwell within your being. I am that tiny light that barely glows inside you, and I am waiting for you to recognize that glow and fan it until it is the roaring flame that you are. Lucifer is running rampant on this planet and trying his best to extinguish that flame. He is about to succeed. Many of you are following him like the Pied Piper of yore, and he is leading you to your destruction.

But you can raise yourself up out of the depths. All you have to do is extend your hand to Me, and I will pull you up out of the pit that so many on this planet abide in.

I have answered your call. My legions of light are here at My command to assist you. But there will be no help until you ask Me, and I will respond only if that request comes from the heart

Many in this third dimensional world say they cannot believe in Me because they cannot see Me. Since the early beginnings on Earth, most have been aware of My presence, and they have gone on faith alone. Many are saying that is not enough. Now, I will not judge those that say that, but I will soon eliminate that situation. I will not reveal how at this time, but at some point in the cleansing process that has already begun I will reveal Myself to all the inhabitants of the planet. This will be an unprecedented event. I have never made Myself visible to any beings in the entire Universe before, not even to the ones I created to help in the creation process.

Original Creator

Much information has been given you in these past several months. You have also brought forth much information from your memory as well. However, as you are quite aware, it is very difficult to keep that information from being distorted. Just the fact that this is coming from a higher dimension alone makes it difficult.

Then add to that the fact that there are forces that will go to any lengths to try to distort the information. Or they might even try to add some of their input into the information that is being transmitted. This has happened to you, sometimes knowingly, sometimes unknowingly. The protection around you has been such that what you have received has been the clearest that has been received by any channel on the planet during the months that you have been a channel. We will continue to make sure that there will be absolutely no interference or disruptions from this point forward.

So we are pleased that you realized that this needed to be done. Also from this point on I will personally "stand guard," as you might put it, during all of the messages as they are sent, to be absolutely sure that there is no interference. If you forget to do your part in the protection, I will give you a gentle reminder. That fact alone will generate much more curiosity of the dark forces, and they will be that much more anxious to penetrate the defenses around Taliesen. You will be totally surprised when you see the activity around your little piece of the being known as planet Earth. It is truly amazing. Of course, the entire planet is like a beehive, but still nothing like your little place that is your temporary home.

Now, as to the reason for My coming to you again after an absence of a few months: I see that you are confused just a bit. I have not really been absent, as I just explained before we started recording this. In the early days while you were still writing and editing your first book, you perceived a difference in the one you now call Creator and the one you called the Ancient One. In reality we are the same energy, the same being that has dropped many levels of existence to be at this lower dimension in order to be a part of the ascension process of this planet and at least some of the ones that live on and within her body. We have been in this lower dimension for a longer period than was comfortable for us; so we felt it necessary to go into a higher dimension for the rest of this experience.

We could only do that after you were able to receive from this higher level of creation. We did not explain to you the reason for that because we knew that we would be communicating with you on a much more frequent basis and could then explain our actions. It is much easier for us from this level. However, we did explain to you that from this time forward you will be the only channel we will use. We know that that will not be an ego problem for you at this time. There will be others who will say that they are still receiving from us, but they will now have to be very observing and question their sources.

You have been notified that the planet and the ones living here at this time are in the final days before the transformation occurs. It is in the beginning stages, and as you have been told, it will now accelerate more than at any time in the past. Lucifer and his close associates are aware of this; so they are spreading the word to the ones of his, shall we say, "army" to begin their final assault to disrupt all the goings-on on the planet. Hence the meetings that you were informed about during the last few days.

You may not be happy about what I will now say because it has been said so many times, but I will be the last to talk of this. We want to be completely assured that you are comfortable with what you will be doing. These will be difficult times for everyone, but you are going to be in a most vulnerable spot as a go-between between us of the higher realms and the hierarchy of the planet. They are going to be very hostile in the beginning, and you will be in a very dangerous situation for a time.

Once they fully realize the situation and who they are dealing with, things will begin to ease up a bit. They will never be happy about what is going to take place; so you will not be able to relax your awareness of what is going on at anytime. Your protection will be total (Narno and Sanat). They will always make their presence known when needed.

With what you are being given and the closeness of the messages, I am sure that you are feeling that the timing is right here and now. Your intuition is correct. We have indicated as much; so you should be prepared for anything at all times. You can expect the planetary shift to happen at any time. Various things have been said to you about that, and I will not talk more about that.

In the beginning we felt that it would be best to bring you aboard for relatively short periods of time and make the transformation over a period of time. You probably think that this could have already happened, but there are a number of reasons that was not possible, which you are now aware of. Every event must be done in its proper time. Everything must be in place. The window is now open; so it can now take place.

One of the main reasons that you were asked to stay on the planet in its present density is because what you have been doing could only have been done in the physical. Now it is because you have agreed to record all that takes place in the coming weeks. This can also be done as you go back and forth between dimensions. Then after a certain event takes place, there will be no reason for you to record any longer.

Michael's Role Continues (Creator)

You would like to know if you are also being prepared for another situation, something taking much preparation. The answer to that question is yes. It has not been the right time to tell you of this in the past, but since the timing is now approaching, and since you have sensed it and asked about it, we will discuss it. Let's go back a few years and get you to remember certain statements you made, seemingly just in jest. A number of times over the years something would prompt you to say, "If I were king, I would" do such and such. Remember?

Now, of course, you are not going to become king for a day or even longer, but know this: after the shift, there is going to be much chaos for quite a while – at least two years. There will be so much to do that it will seem like you need to be in several places at once. That is why Joshua is here. When I asked him to be part of this, I indicated I would like him to be your second in command. I also told him it would be up to you as to what part he would play in this.

Bobby's First Book (St. Germain)

Okay, about your book... We have only given you a little information about the procedure that is involved. We have told you that it would be placed in this incredible word processor computer that has already been programmed to take your disk and go through an editing process. There will be some rearranging and comparing it with certain facts to make sure that it is completely in line with what is to be given to the population of the planet. You are, of course, aware that it was to go through this process and that you are comfortable with it.

Now, this computer is tied in with, for want of a better phrase, the complete history of all that has taken place in the universe since before the beginning. How's that for a mind-boggling fact! Who would have thought it possible that our Creator would allow such as that to happen? But He is the instigator of updating this information, to completely modernize the Akashic records. So it's just a matter of punching in what you are looking for, and the information is then brought forth in a matter of minutes, depending on what you are looking for and how extensive the questions are.

Yes, Bobby, to what you are thinking. When you were writing your book and getting the information that you were searching for, that was what was happening. You were the first to use the computer, even though you were not here in the same presence with it. It was simply attuned to your energy, and it responded to your questions by way of your thoughts, even though you were not aware that they were questions. Now, as you can imagine, the security is enormous. Only ones that are authorized and have the correct energy pattern can gain entrance. How awesome is that?

When you were told that a new system was in operation, little did you realize that it would be like this. What you were asked to do was to reduce that total amount of information into a small book size – a book that was easy to read by most of the inhabitants. So what will be done by submitting it to the computer is to compare your facts to the information and make any changes of facts and order as needed. So you can see the necessity of writing what you did and the necessity of comparing it with the total amount. As you have already surmised, this is a living conscious being that we are describing as a computer.

When it is complete, through a process that can only be accomplished by our Creator, the contents of the book will be programmed into the consciousness of all of the inhabitants of the planet. That is the only way that it can be accomplished. Now, we know that there will be many that will not be receptive to this. We will also make the book available in a visible format that will be unlike any book that has been published and made available to readers.

www.trafford.com

North America & international
toll-free: 1 888 232 4444 (USA & Canada)
phone: 250 383 6864 ♦ fax: 250 383 6804
email: info@trafford.com

The United Kingdom & Europe
phone: +44 (0)1865 487 395 ♦ local rate: 0845 230 9601
facsimile: +44 (0)1865 481 507 ♦ email: info.uk@trafford.com